Enjoy This Book!

Manage your library account and
discover all we offer by visiting us
online at www.nashualibrary.org

Please return t
so others ca

If you love
library offers, tell a friend!

@ Nashua Public Library
2 Court Street, Nashua, NH 03060
603-589-4600, www.nashualibrary.org

DRIVE LIKE THE PROS

GOTHAM BOOKS

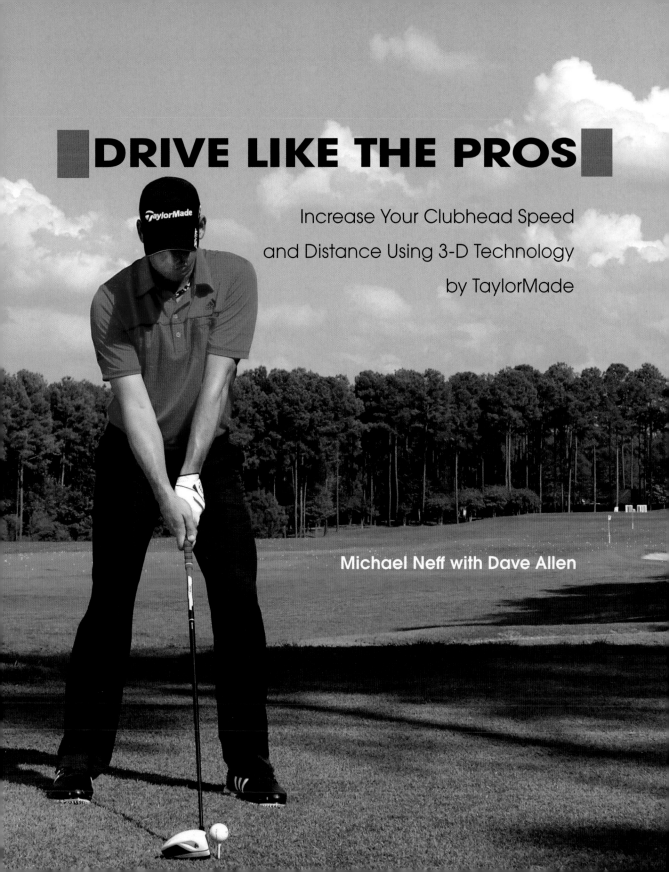

DRIVE LIKE THE PROS

Increase Your Clubhead Speed

and Distance Using 3-D Technology

by TaylorMade

Michael Neff with Dave Allen

GOTHAM BOOKS
Published by the Penguin Group
Penguin Group (USA) Inc., 375 Hudson Street,
New York, New York 10014, USA

USA | Canada | UK | Ireland | Australia | New Zealand | India | South Africa | China
Penguin Books Ltd, Registered Offices: 80 Strand, London WC2R 0RL, England
For more information about the Penguin Group visit penguin.com.

LIBRARY OF CONGRESS CATALOGING-IN-PUBLICATION DATA
Neff, Michael.
 Drive like the pros : increase your clubhead speed and distance using 3-d technology by Taylormade
/ by Michael Neff, with Dave Allen.
  p. cm.
 ISBN 978-1-59240-727-9
1. Swing (Golf) 2. Three-dimensional imaging. I. Allen, Dave. II. Title.
 GV979.S9N434 2013

 796.352'3—dc23   2012029939

Printed in the United States of America

10 9 8 7 6 5 4 3 2 1

Set in Adobe Caslon Pro
Designed by Patrice Sheridan

would like to dedicate this book to my eternal family—

my "really patient" wife, Janea, and my children,

Samuel, Gabriel, Henry, Charlotte, and Lucy.

I love you so much and pray I can support you

in all the great things you desire to achieve in life.

CONTENTS

FOREWORD

AS A CORPORATE CONSULTANT for TaylorMade-adidas Golf (TMaG), I was first introduced to the Motion Analysis Technology by Taylor-Made (MAT-T) system at The Kingdom in Carlsbad, Calif., some seven or eight years ago. What struck me right away was just how sharp each image was, and how many angles there were. You could clearly see the shaft plane and the position of the club and body from several different angles without the usual distortion you get from using video cameras. I was equally captivated by all the statistics that were popping up on the screen in front of me. I wasn't just seeing launch numbers, I was also able to tell how open or closed the clubface was at impact, what the path of the clubhead was, and whether there was too much or too little side bend and forward tilt at impact. There were all sorts of numbers to analyze, and I was eager to learn more about the system and how I might use it to help my students.

Unlike most swing analysis programs, the MAT-T system helps us to understand where the problem really lies. It doesn't mislead you into fixing the wrong things. For example: A typical launch monitor system measures ball speed and spin rate, and then extrapolates clubhead speed from these numbers. In other words, it guesses, and in the process it ignores such influential factors as clubhead path, attack angle, where the clubhead impacts the ball (i.e., toward the toe or heel), etc. Therefore, when a player with 100 mph clubhead speed shows up for a fitting and discovers that his ball speed is only 130 mph, he automatically thinks that he needs more speed to get the maximum performance out of his driver. He doesn't stop to consider that he needs to become more efficient with the speed he already has. The MAT-T system helps you to identify where your swing's inefficiencies are, so that you can correct them and build a more efficient, powerful motion without having to swing out of your shoes.

That's the key to hitting the ball longer—efficiency of motion and efficiency in how you're using the club. The MAT-T system allows you to take a look at yourself and see from a motion and sequencing perspective what's going on with your swing, so there's no guesswork. Once you pinpoint where the motion or sequencing breaks down, you can determine if there's a physical explanation or an equipment or mechanical cause to the problem. The 3-D motion capture technology eliminates any confusion there might be with regard to what you're trying to do with your golf swing.

For me, the MAT-T system has substantiated a lot of the things that I already knew about the swing, especially as it pertains to the physical side. I've been a huge advocate over the years for more fitness education in golf, and I feel like now, with the help of such diagnostic tools as the MAT-T system, people are finally beginning to understand just how important their overall fitness level is to their performance. There are many different physical limitations that can cause you to swing the club off-plane or to lose clubhead speed, and the MAT-T system can help you determine where these weaknesses are.

Once you gain a better understanding of who you are (your physical makeup, emotional characteristics, swing tendencies) and how you move (body sequencing), you'll be better able to determine the right path of improvement for you. This book is a great place to start, and I would also encourage you to visit one of TaylorMade's Fitting Centers or Performance Labs (see taylormadegolf.com for locations) so that you can experience the MAT-T system for yourself. If you want to hit the ball farther, there's no better resource from both a clubfitting and swing assessment standpoint. Two hours on the MAT-T system will make you a much better player, without question.

—MIKE MALASKA, 2011 PGA TEACHER OF THE YEAR

INTRODUCTION

Pass the 3-D Glasses and Popcorn

TWENTY YEARS AGO, THE longest driver on the PGA Tour averaged 280 yards off the tee. Today, that would be considered short by Tour standards. Consider: As of the 2012 PGA Championship, 170 players averaged at least that much off the tee. The average drive was almost 290 yards, with twenty-two players boasting averages of more than 300 yards per poke. At the 2012 Hyundai Tournament of Champions, one of those nearly two dozen players, Gary Woodland, hit a tape-measure shot normally reserved for a long drive championship. His ball traveled 450 yards, or approximately three-tenths of a mile.

Professional golfers really are driving for show, and for dough, too. Six of the top seventeen players in Total Money earnings on the Tour in 2011 also ranked in the top twenty in Driving Distance. The enormous increase in distance over the past twenty years can be linked to several different factors, most notably better fitness and improved equipment and technology. Persimmon woods and one-piece golf balls have been replaced by stealthy 460cc titanium heads and multilayer, urethane-covered balls that launch off the clubface like a rocket. Players can now adjust their drivers with a few turns of a wrench to optimize the ball flight they desire. Equipment truly has become personalized.

The increasing use of video, swing analysis software, and launch monitor systems in instruction has also played a major role in fueling the latest power surge. A player can see exactly what's happening during his swing and then use the information to maximize launch conditions (e.g., high launch, low spin) and better fit their equipment to their swing. The most state-of-the-art system today is the TaylorMade MAT-T System. Originally designed for clubfitting purposes, the MAT-T system combines multiple high-speed cameras, a launch monitor, and specifically designed software to produce a three-dimensional, computer-animated image of the golfer's swing that's viewable from virtually any angle. It analyzes every aspect of the golf swing, from the angle of your shoulders at the top of the swing to the cocking angle (i.e., the angle between your left forearm and the clubshaft) on the downswing to the angle of descent the clubhead takes into the ball, performing a multitude of calculations and measurements that make it an exceptionally useful swing aid.

The advantages to using 3-D motion analysis over 2-D are enormous. With the MAT-T system, a certified golf professional is able to take a golfer and measure to a tenth of a degree what the body and club are doing during the entire swing. He can calculate the exact angle of the shoulders, hips, knees, feet, and spine of the golfer at every position in the golf swing, and show how the body moves from one position to the next from a multitude of angles—above, face-on, down the line, and from behind. As a comparison tool, he can also bring up a composite golf swing of more than a hundred Tour professionals and all of their swing DNA, including the likes of Dustin Johnson (third on the PGA Tour in driving distance in 2011 at 314.2 yards), Martin Laird (303 yards), Jason Day (302.6), and Sergio Garcia (299.2). The MAT-T system allows the instructor to overlay your own personal swing avatar (i.e., an animated computer image) over the Tour composite model, so that you can compare your golf swing to the composite swing or to that of an individual avatar of one of these players. This allows you to feel exactly what the Tour player feels, and to see how the game's longest hitters align themselves and move their bodies throughout the swing. Your biggest swing flaws will be made evident, so that you can take corrective measures to improve your swing and take your game to the next level.

This live comparison tool is the holy grail of golf instruction, and the foundation of *Drive Like the Pros*. Using this composite swing along with the individual models of such players as Johnson, Garcia, two-time U.S. Open champion Retief Goosen and Sean O'Hair, you'll see how the typical over-the-top slicer stacks up against the very best players in the world, position by position, and in the process learn how to plug all of those power leaks in your swing so that you can reach your full power potential. You'll learn all the Tour pros' secrets to hitting the ball longer and straighter, from how to align your shoulders and where to play the ball at address, to how to widen the differential between your hips and shoulders at impact (the "Torque Factor") so you can make your body turn faster through the ball.

No other golf instruction book on the market today can detail, frame by frame, the exact movements of the club and body of the game's longest hitters like *Drive Like the Pros* can. By reading *Drive Like the Pros* and working on the individual drills and lessons in this book, you'll be able to accomplish a number of different things you may have previously thought were impossible. For example, you'll learn how to:

- Increase your clubhead speed, ball speed, and Smash Factor (efficiency of contact), and hit longer drives; some students I've worked with have gained as much as 50 extra yards utilizing the information they received from the MAT-T system.
- Contact the sweet spot on the clubface more consistently to keep your longest drives in play.
- Make a healthier, more efficient swing so that you can avoid injuries and enjoy the game for years to come.
- Lower your scores and, just maybe, outdrive everyone in your weekend foursome.

There are forty MAT-T systems currently in existence, and many more on the way. The data collected from these systems (currently about twenty thousand swings per year, including the swings of Tour players, 5-handicappers, 15-handicappers, and over-the-top slicers) has never been published before . . . until now. With *Drive Like the Pros,* you'll learn the same valuable lessons that Tour pro Brian Henninger used to increase his clubhead speed from 102 to 109 mph—in a matter of days. That translates to 25 extra yards! Now who couldn't use that?

DRIVE LIKE THE PROS

figure 1-1

THE GOLF SWING'S MRI

MOST OF YOU READING this book have probably undergone an MRI (Magnetic Resonance Imaging) exam at some point in your lives. Perhaps you hurt your knee playing basketball, or experienced some sharp neck or lower back discomfort when playing golf. You went to see a specialist for answers, and he had you lie under this giant magnetic tube that looks more suitable for space travel than a doctor's suite.

Using radio frequency waves, this MRI scanner produces a rotating magnetic field that allows a radiologist to view important internal anatomy, in particular soft-tissue structures (muscles, ligaments, tendons, vital organs) that cannot be seen as well with other imaging techniques—such as a CT scan or traditional X-ray. What's more, the MRI image can be viewed on several different planes, three-dimensionally, making it much more comprehensive than most imaging tests. After the MRI is complete, the radiologist interprets the results, sends a detailed report to your doctor, and you usually have an answer within twenty-four to seventy-two hours.

The MRI is the "gold standard" of diagnostic tests for your health, just as the MAT-T system is for your golf swing and equipment needs. Co-developed by Motion Reality Inc. (MRI) and TaylorMade in 2001, the MAT-T system (MAT-T stands for Motion Analysis Technology by TaylorMade) was created to provide golfers with a three-dimensional clubfitting experience like nothing else. But over time, it's also become

■ 1

known as an exceptional instructional tool. Like an MRI, the system looks at an object from three dimensions, and then attaches math to what it's measuring. The motion capture software triangulates the image of the golfer so that the swing can be stopped at any point and viewed from above, below, straight on, down the line, etc. A launch monitor is also used to record ball data. This allows the clubfitter or pro (think of them as your radiologist or doctor) to measure exactly what your club and body are doing at every point in the swing.

The latter is what separates the MAT-T system from traditional launch monitors and ball-flight-analysis systems. It provides detailed information about what your club and body are doing, not just the ball. Most launch monitors use the ball-flight data (spin rate, launch angle, ball speed, etc.) to guess what the club is doing (i.e., how much loft is being presented to the ball, the up-and-down path, face angle, clubhead speed, etc.). They're not able to track the club precisely. This is an important distinction because the club tells the ball what to do, so it's really important that we understand what it is the club is doing at impact so that we can have greater control over where the ball is going.

If, for example, it was determined that your last tee shot was launched at 15° with 140 mph ball speed, 3,000 rpm backspin, and 300 rpm side-spin, the MAT-T system can tell you how the club created these numbers. In other words, it's able to tell you the exact angle of the clubface at impact (within one-tenth of a degree), and both the direction (or path) and angle (up or down) that the clubhead took into the ball. With the MAT-T system, you're also able to see what the body and club were doing well before and after the clubhead entered the hitting area, which has a great influence on your impact position and how far and straight you hit the ball. The MAT-T system tells us the "why," while most launch monitor systems only tell us the "what."

■ HOW IT EVOLVED

The idea of being able to measure what the club is doing (for R&D and clubfitting purposes) is precisely why TaylorMade approached Motion

Reality Inc. to help them develop the MAT-T system. A pioneer in the area of computer graphics animation, MRI won an Academy Award for Technical Achievement in 2005 for their groundbreaking work in all three *Lord of the Rings* films, and other films such as *The Polar Express* and *King Kong*. MRI is able to capture a subject's movement in 3-D and not only bring it to life on-screen, but also present important bio-mechanical data associated with the motion. It is able to measure things—such as the angle of your spine, hips, and shoulders throughout the entire swinging motion—that couldn't be measured ten years ago. That's what makes the MAT-T system so effective: By looking at your computer-animated image and the data associated with your swing, I can tell you with certainty if you're swinging over the top or if you're coming up out of your spine angle, and exactly when it's occurring.

I became certified as a MAT-T specialist in 2002. At the time, there was only one system in use, and that was at The Kingdom—a state-of-the-art clubfitting facility for TaylorMade Tour Staff Professionals in Carlsbad, Calif. TaylorMade had invested millions of dollars in the MAT-T system, and was at a crossroads as to what to do with it—either continue to use it internally for research and development (R&D) purposes, or outsource it and make it available to the public. The manufacturer chose to do both, and today there are as many as forty systems in place worldwide, with many more than a hundred expected to be fully operational by 2014.

All of the TaylorMade Performance Lab locations, including the one at The Kingdom, utilize the MAT-T system. Golfers can book a two-hour fitting session with a certified TaylorMade clubfitting professional at one of these locations, in which they'll undergo a complete analysis of all their clubs on the MAT-T system. Not only will they be able to view their swing in 3-D at the lab and on their own personal computer, but they'll also be able to see how their swing stacks up against some of the game's biggest hitters, including Dustin Johnson, Sergio Garcia, and Jason Day. More than a hundred TaylorMade Staff Tour Professionals have been on the MAT-T system, and all of their swings are viewable, as is a Tour composite model of all their swings (see Chapter 2 for a detailed analysis of this composite swing).

In the U.S. alone, more than twenty thousand swings were captured and stored on the MAT-T system in 2011. These include swings of Tour players, single-digit handicaps, 15-handicaps, and slicers. For the purposes of this book, we've averaged together all of the swings of the Tour pros and thousands of over-the-top slicers, so that you can clearly see the differences between a 300-yard driver swing and a 200-yard driver swing. There is no better comparison tool in golf, and after reading this book I guarantee you that you'll know why it is you're losing yardage and, more importantly, how you can get it back!

■ HOW IT WORKS

If you schedule a clubfitting on the MAT-T system, the first thing the certified fitter/instructor will do is ask you about your goals, habits, ball-

figure 1-2

flight tendencies, prior injuries, etc. This interview lasts approximately five minutes. Then the fitter will attach thirty-eight reflective markers to your body and club. Six of these markers are on the club's head and grip, nine on your torso, four around your waist, three on your head (or whatever hat you're wearing), three on each foot, one on each hip, two on each knee, one on each elbow, and one on each wrist. A thin vest is placed around your torso and noninvasive neoprene sleeves are applied to your feet, knees, waist, elbows, and wrists to help secure the markers to your body. No suit is required; you simply show up for the session in your normal golf attire. Each marker is attached with Velcro, so the markers don't move around and inhibit your swing.

The purpose behind the markers is twofold: 1) To allow the cameras to pick up the motion of your body and club; and 2) to help create a computer animated avatar of your body. The fitter enters your height and waist size into the computer, then goes through a process of using the

figure 1-3

markers to fill in the rest of your body. Each marker is strategically positioned so that the computer knows the distance between them and can calculate how long your arms, legs, torso, etc. are. A "skin" is then laid over your body and the club, bringing your swing and all its moving parts to life.

Six high-speed cameras with high-powered LED lights are positioned around the room to pick up the motion of all thirty-eight reflective markers and convey the information gathered from each swing to the computer. The lights reflect brightly on the markers, making it very easy for the computer to single them out. The cameras make it possible to view your computer-animated swing from as many as six angles, including the traditional face-on and down-the-line aspects, along with views from above, below, behind, and up the line (i.e., from the target).

Once the markers are in place, the fitter will have you make five or six swings with your driver, 6-iron, and sand wedge, and he'll then go over all of the pertinent ball, club, and body data collected by the computer. (The golf professional also utilizes a launch monitor to measure what the ball is doing.) The MAT-T system uses a very smart diagnostic engine to recommend the clubs you should test. The golf professional then prepares the clubs and has you hit them outside on a grass driving range to validate the club's performance—and view the ball flight.

■ THE MAT-T SYSTEM AS A TEACHER

As master clubfitter and PGA teaching professional at Pumpkin Ridge Golf Club in Portland, Ore., I'm also in charge of the training and installation of the MAT-T system worldwide for TaylorMade. I have trained more than 200 golf professionals to use the system, and have seen more swings and statistics on the MAT-T than anyone else on the planet.

When I'm teaching in Oregon, I like to use the system for both clubfitting and instructional purposes. It depends on what the student wants. The ability to measure and capture what the body is doing throughout the swing makes it a tremendous teaching aid. Perhaps its greatest asset, though, is as a visual/feedback tool, because it can freeze any position in the swing and look at it from six different angles, or 360°. A regular 2-D video camera can't do this; all you typically see is your swing from either a face-on or down-the-line view. The motion capture system triangulates your image so that you can even view your swing from below, even though there's no camera below the floor. This is particularly helpful when you want to see the movement of your feet.

The MAT-T system has definitely changed the way I teach the golf swing. I'm now able to explain things faster to my students, and they get it quicker. The learning curve isn't as steep. The Tour composite comparison tool plays a big part in this: It allows them to see what they're doing wrong (compared to a Tour pro). I'm then able to move them into the correct position so that they can immediately feel and see where they need to be.

Every student I've taught on the MAT-T system has picked up additional yardage off the tee. I had one fifty-two-year-old man gain 50 yards in a single day after going to a lighter, more flexible shaft and learning how to hinge his wrists properly on the backswing. He increased his clubhead speed and ball speed substantially, adding 35 yards of carry and 15 yards of roll. One session was all he needed, because we were able to identify what his biggest power leaks were and fix them. That's what this book will do for you: It will give you a better understanding of what the most common power faults are in the swing, and what causes them. It will also provide you with a road map—thanks to the Tour composite swing and the data attached to this swing—to a stronger, more powerful game.

To find the TaylorMade Performance Lab and MAT-T system nearest you, go to taylormadegolf.com.

figure 2-1

Sergio Garcia's swing is one of more than 100 that make up the MAT-T system's Tour composite swing.

THE TOUR COMPOSITE SWING

SINCE 2003, MORE THAN a hundred TaylorMade staff pros, from major championship winners Nick Faldo, Retief Goosen, Darren Clarke, and Martin Kaymer to bombers like Dustin Johnson, Jason Day, Sergio Garcia, and Sean O'Hair, have had their swings captured on the MAT-T system. What follows is a composite golf swing sequence of all these swings—the ultimate playlist, if you will, of the most efficient, powerful swings in the game.

Every time I give a student a lesson on the MAT-T system, I put their swing avatar up against this composite model, which allows the student to see precisely where his club and body are in comparison to the Tour model at every key position in the swing. It's the gold standard as far as swing sequences go, because it's not just one Tour player's swing but a composite of 118 of the best swings in the world. What's more, the sequence can be viewed from any angle (most swing sequences you see in magazines and books show only one or two), and each frame comes with actual numerical data (body lines, joint angles, club positions, etc.) that, until now, was only available to a few Tour pros, instructors, and clubfitters.

In this chapter, I'll break down the composite golf swing of these players, with a statistical blow-by-blow account of the nine most commonly analyzed positions in the golf swing. By the time you're done reading this chapter, you'll know just how much the Tour player turns his shoulders on the backswing, how much lag his club has at the start of the downswing, and how far his hips actually rotate prior to impact. You'll also have a much better understanding of the angles that the Tour player creates during the swing, and how he uses these to generate maximum clubhead speed and power.

Some of the most commonly asked questions I get from students, such as, "Where should my shoulders point at address?" or "What starts the downswing?," will all be answered by this composite sequence. No two Tour players swing the club in exactly the same way, but they all have certain moves in common that are evident in this composite model. What you'll see on the following pages is as close to an "authentic" Tour swing as you'll find anywhere.

POSITION 1: ADDRESS

The average Tour pro has 14.4° of shoulder tilt (represented by the red line) and 3.7° of spine tilt (blue) with the driver [figure 2-2], compared to 11.7° and 2.9° with a 6-iron [figure 2-3].

The average tour player hits slightly up on the ball with the driver (3°), because the ball sits on a tee. This makes the driver unique, since every other club in the bag requires that you hit down on the ball. To create this ascending path, the tour pro makes a few minor adjustments to his setup, mainly widening his stance and moving the ball more forward,

figure 2-2

figure 2-3

opposite the left armpit. This forward ball position, coupled with the fact that the right hand sits lower on the shaft than the left, puts the shoulders on a 14.4° incline away from the target [figure 2-2]. The spine naturally leans back—3.7° compared to 2.9° with a 6-iron [figure 2-3]—while the hips remain level. There's a slight lean (away from the target) to the shaft as well, and this puts the hands in line with the left thigh and the grip slightly behind the ball. The stance is wide enough so that the feet are shoulder-width apart—as measured from the insides of the heels.

This is the most inclined that the shoulders and spine will be for any club in the bag. The spine is slightly tilted away from the target, which puts our Tour player in the best position to make a slightly ascending blow on the ball (3°)—provided he maintains this side tilt and his forward tilt toward the ground (29.9°) through impact [figure 2-4]. With a shorter club,

Due to the length of the driver, the spine is inclined less toward the ground (29.9°) than it is with a 6-iron (34.9°), and the feet are farther away from the ball [figures 2-4 and 2-5].

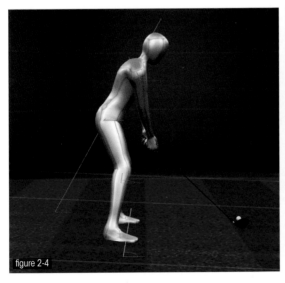

figure 2-4

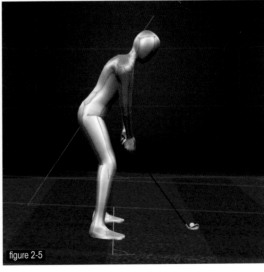

figure 2-5

such as a 6-iron, the spine will be inclined more toward the ground (34.9°) and the feet will be much closer to the ball [figure 2-5], but because the driver is also the longest club in the bag, our composite golfer is standing farther away to accommodate the length. His hands are directly under his chin and farther away from his body than they are with an iron or wedge, and there's a straight line running through the back of his shoulder blades, kneecaps, and the balls of his feet. In this athletic posture, he's able to turn around his spine and swing the club on a circle, which is what produces angular momentum and maximum clubhead speed.

■ ALIGNMENT: FEET SQUARE, SHOULDERS LEFT

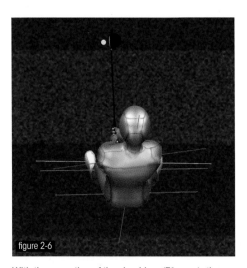

figure 2-6

With the exception of the shoulders (7° open), the average Tour pro sets up with his toes (yellow line), knees (purple), and hips (green) relatively square to the target line.

From above you can clearly see that our composite tour model's shoulders are pointing left (7.0° open) of the target line, while the hips (1.3°), knees (0.7°), and toes (-0.9° closed) are relatively square to the target line [figure 2-6]. Only the toes point right of the target line, albeit a minute amount. Our Tour pro is setting up to hit a baby fade—a ball that starts just left of the target line and moves back toward the line. This shot will look very straight to the average golfer. The baby fade is easier to control than a draw (a shot that starts right of the target line and curves back toward the line) because the ball tends to land softer with less roll, although it doesn't generate as much distance.

■ POSITION 2: MID-BACK (SHAFT PARALLEL TO GROUND)

What's interesting to note about this frame is the movement of the shoulders. Whereas the shoulders at address were inclined away from the

Shoulder tilt: -23.8° (toward target)

Shoulder rotation: -41.8°

Hip rotation: -24.4°

target 14.4°, at mid-back they're tilted toward the target -23.8°, a difference of more than 38°. This tells us that during the first few feet of the swing, the Tour player's left shoulder drops down from its original address position [figure 2-7]. The left hip and knee also work downward, toward the ground, while the right knee and hip extend slightly upward. The reason this happens is that it's the only way to keep the spine in its original inclination. From the address position, neither spine angle (forward or side tilts) has changed more than 1.5°. This is something all good players do: As they turn back, their spine angle hardly moves; it stays in its original inclination with very little up-and-down or side-to-side bending [figure 2-8]. It is what allows them to swing the shaft on-plane—in a circle at right angles to the spine—which is the fastest, most efficient way to swing a club.

From above, we can see that both the hips and shoulders are turning as the left arm swings across the player's chest [figure 2-9]. The shoulders are rotating almost twice as fast as the hips (-41.8° to -24.4°), and they will continue to out-turn the hips as the arms swing to the top. The wrists do begin to hinge slightly, but the back of the left wrist remains flat; it doesn't cup—i.e., bend back toward

figure 2-7

During the first few feet of the swing, the left shoulder drops down while the right knee and hip extend slightly upward.

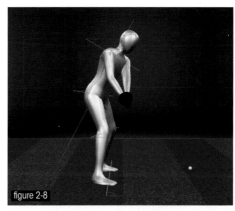

figure 2-8

The forward spine tilt (28.6°), or inclination, remains relatively unchanged from the address position.

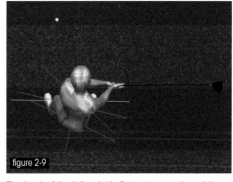

figure 2-9

The back of the left wrist is flat, not cupped, and the clubhead is directly in line with the hands, not inside or outside the hands.

the forearm, thus opening the clubface. The clubhead is directly in line with the hands, and the angle of the face nearly matches that of the spine. From a down-the-line view, you can also see that the composite player's left knee moves forward, toward the ball. Again, this helps the player maintain his original spine inclination as he swings back.

POSITION 3: LEFT ARM PARALLEL

Left: The angle formed by the shaft and left arm—also known as the cocking angle (81.1° in this image)—is a key to producing more power.
Right: The left side continues to drop down, which helps the golfer remain in his original spine inclination (still 28.6°).

By the time the composite model's hands reach nine o'clock on the backswing, the wrists are almost done hinging. The cocking angle—the angle formed by the shaft and left forearm [figure 2-10]—is 81.1°, creating an additional lever that, when released by the unhinging of the wrists, multiplies the speed of the clubhead through impact.

The left shoulder, hip, and knee continue to drop down toward the ground as the right side extends. Again, this is the only way for the spine to remain in its original inclination (i.e., forward and side tilts). The left arm is extended (6° left elbow bend), but not locked, a sign of excellent width—

figure 2-10

figure 2-11

the farther you swing the butt of the club away from your body without losing your spine's inclination, the more clubhead speed you'll generate.

From down the line, note the position of the player's hands directly in front of his right biceps [figure 2-11]. This tells us that the hands are on-plane (more on this shortly) and that the player has done a good job of synchronizing his arm swing with his body turn. The left arm is swinging across the Tour player's chest, and the hands have moved slightly in and up in response to the turning of the shoulders. At this point, the player has already turned his shoulders 87.7° and his hips 40.3° [figure 2-12]. Neither has much farther to go, and yet both spine tilts remain relatively unchanged from address. It takes a great deal of flexibility, but the farther you can turn your shoulders without losing your spine angle, the farther you're going to hit the ball.

figure 2-12

From above, you can see that the shoulders (-87.7°) are already out-rotating the hips (-40.3°) by a more than 2-to-1 margin.

POSITION 4: TOP OF BACKSWING

The tip of the left shoulder on our composite player points to the right-center of his stance, almost to his right knee, an indication of just how much he's turned his shoulders [figure 2-13]. The shoulders (-122.6°) have out-turned the hips (-48.8°) by a nearly 3:1 ratio, creating quite an X-Factor [figure 2-14]—the term invented by noted instructor Jim McLean to describe the difference between shoulder rotation and hip rotation at the top of the swing. (The greater the differential, the more clubhead speed you'll generate on the downswing.) But what's most remarkable about these numbers is that they've been accomplished with minimal alterations to the spine; in fact, the side tilt of the spine has increased only five-tenths of a degree, while the forward tilt has moved just .9° toward the ground. This is critical, because it doesn't matter how much you turn your shoulders and hips if your spine doesn't remain in its original inclination. The X-Factor is only a strong indicator of your power potential if your spine angle remains intact.

Left: The left shoulder turns down and across the chest and points behind the ball at the top of the backswing.
Right: The differential in shoulder rotation (-122.6°) to hip rotation (-48.8°) sets up a powerful uncoiling of the body on the downswing.

figure 2-13

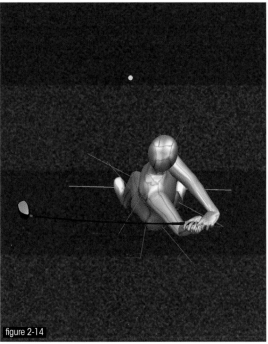

figure 2-14

The other thing to note about this frame is the position of the left arm, which is on the same line as the shoulders [figure 2-15]. The hands are directly over the right biceps and the clubface is perfectly square, matching the angle of the left forearm. From here, our composite player is in excellent position to start the downswing—his hands and club are on-plane, the face is square, and he's stored a tremendous amount of power. In addition, he's already starting to shift his center of gravity forward, onto his front side. This increases the club's cocking angle to 112°, creating more lag between the clubhead and the hands and setting up a powerful release of energy through the ball.

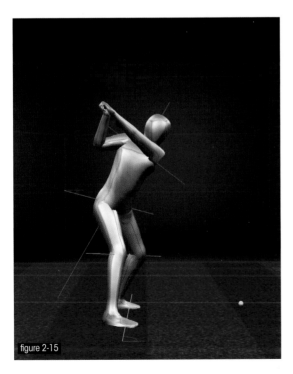

figure 2-15

Our Tour composite golfer's spine angle (30.8°) has hardly changed from address despite the enormous amount of shoulder and hip rotation.

■ POSITION 5: TRANSITION (LEFT ARM PARALLEL)

As the Tour player makes the transition from backswing to downswing, his left knee and hip start to move forward, toward the target, and up [figure 2-16]. In fact, the left hip moves up quite significantly (13.1°) from the start of the transition until the time the left arm is parallel to the ground. As the left side extends upward, the right side goes down, the complete opposite of what occurs at the start of the swing.

While the lower body nudges forward, the spine remains in its original inclination and the hips and shoulders begin to unwind [figure 2-17]. Again, the shoulders unwind, or rotate, at a faster rate than the hips (67.7° to 46.7° from the top), but that's only because they have a longer distance to travel. The hips (-2.1° closed) are clearing fast, and they've nearly returned to their original address position (1.3° open). There's still a fairly large separation of more than 50° in the amount of hip and shoulder rotation, meaning that there's a lot of stored energy waiting to

Left: At the start of the downswing, the roles are reversed and the right side begins to drop while the left side extends upward. Right: The hips remain slightly closed to the target (-2.1°) while the shoulders (-54.9°) point well right of the target.

figure 2-16

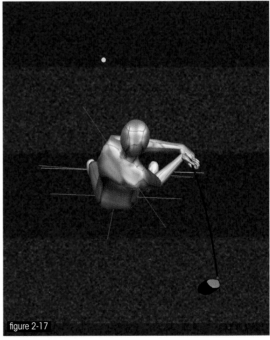

figure 2-17

be unleashed through the ball. The cocking angle (107°) also remains virtually unchanged from the top of the swing, adding further fuel to the clubhead.

From down the line, you can see that the player's hands have returned to the same position they were when the left arm was parallel to the ground on the backswing and in front of the right biceps [figure 2-18]. The hands are moving straight down, not out toward the target line, which allows the player to maintain his spine angle and swing the club down on-plane, at a right angle to his spine. Again, this is the fastest, most efficient way to swing a golf club, and is a big reason why Tour players are able to routinely drive the ball 300 yards or more, and straight.

figure 2-18

Again, the spine angle (30.6°) remains virtually unchanged from address while the hands return on-plane, opposite the right biceps.

In order for the clubhead to avoid hitting the ground, the spine angle (forward tilt) starts to come up slightly—a 5.3° increase from the previous frame [figure 2-19]. It also starts to back up—a 5.6° increase in spine angle from the left arm parallel position—a result of the composite golfer's hips pushing forward and up [figure 2-20]. The golfer's center of gravity has moved forward of center, in line with his left pec muscle, which also contributes to the increase in spine tilt away from the target. The hands are over his right thigh and moving in a straight line toward the ball, which is the fastest route they can take into the ball.

From above, you can see that the hips are now open to the target line while the shoulders are still playing catch-up [figure 2-21]. The gap between the two has narrowed slightly from the previous frame, but it's still fairly substantial. A down-the-line view shows the left arm pointing to a spot on the ground between the toes and the ball, but closer to the

Left: The player's forward spine angle (25.3°) comes up several degrees to prevent the clubhead from striking the ground.
Right: His spine also starts to back up (9.8°) as a result of his hips and center of gravity pushing forward, toward the ball.

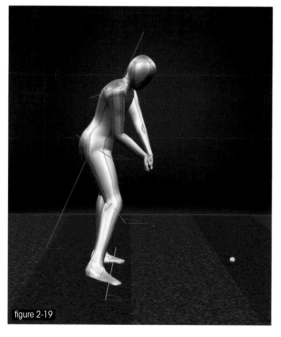

figure 2-19

figure 2-20

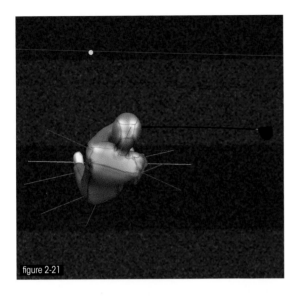
figure 2-21

The shoulders (-24.3°) still remain closed to the target, although the gap between them and the hips (20.9°) is shrinking.

feet. This tells us that the shaft is on-plane and the clubhead is approaching the ball from the inside, which is the optimal path it can take as it allows for a full release and maximum clubhead speed.

KEY MEASUREMENTS

Forward spine tilt: 25.3°

Side spine tilt: 9.8°

Shoulders-hip differential: 45.2°

POSITION 7: IMPACT

At the moment of truth, the shoulders are just beginning to square (-1.1° closed) while the hips continue to open (36.1°). The shoulders will eventually catch the hips, but at the point of contact, the composite player has a substantial impact X-Factor of 37.2° [figure 2-22]. As you'll see later, players such as Dustin Johnson and Sergio Garcia have an even greater separation between the shoulders and hips at impact. This is significant because the wider the gap, the farther you'll hit the ball—provided your hands remain on-plane, which they are in this Tour composite swing

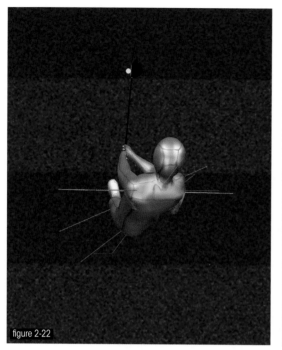

figure 2-22

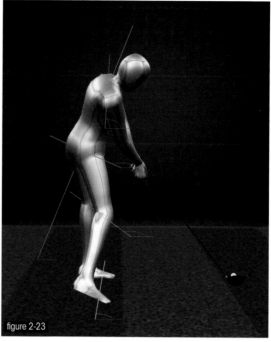

figure 2-23

Left: At impact, the shoulders (-1.1° closed) are just beginning to square up. The differential between the shoulders and hips (36.1° open) is still a healthy 37.2°.
Right: The forward spine tilt (23.1°) continues to rise while the lead hip clears out of the way to make room for the arms.

sequence. From the beginning of the transition, the player's shoulders have rotated 121.5° and the hips 84.9°. The clearing of the hips paves the way for the arms to swing down in front of the body [figure 2-23], and it also helps create the separation between the hips and shoulders, which propels the clubhead through as though it were being fired by a slingshot.

Our composite golfer's spine angle has backed up even farther, to 15.3°, once again due to his hips pushing forward and up, while his center of gravity has moved forward (it's now off the left shoulder). His forward spine tilt has also increased 2.2° from the previous frame, which

KEY MEASUREMENTS

Shoulders-hip differential: 37.2°

Right wrist angle: 23°

Side spine tilt: 15.3°

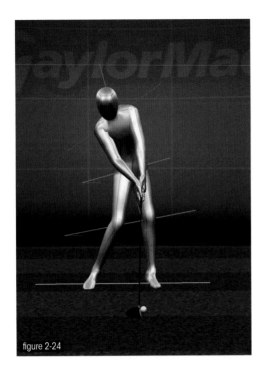

figure 2-24

is designed to prevent the clubhead from bottoming out too soon and hitting the ground. The left shoulder, hip, and knee are all ascending, as is the clubhead, which contacts the ball on a 3° up path. Our golfer's hands are on top of the ball, in front of his left leg, and the shaft is leaning slightly toward the target [figure 2-24]. So while he's hitting up on the ball, he's not adding any loft to the club—this is how you hit a ball with the optimal combination of high launch and low spin. The back of his right wrist is angled at 23° and he still has yet to fully unhinge his wrists; as a result, the clubhead will not reach its maximum speed until the moment of impact.

The spine is really tilted back now (15.3°) and the left side has all but straightened, which allows him to hit up on the ball 3°.

POSITION 8: MID-THROUGH

The shoulders have finally caught the hips, and both are now rotating at the same rate of speed. But the telltale image of this frame is the full extension of the right arm and clubshaft: The clubhead is farther away from his body than at any point during the swing [figure 2-25]. There's very little bend to the left elbow (18.9°) and minimal separation between the elbows (i.e., no chicken wing). What's more, the spine remains relatively close to its original inclination at address, the biggest change being an 8.5° increase in side bend. This is largely due to the centrifugal force created by the clubhead swinging out away from the body. Keeping the spine in this side bend position is what allows the club to stay on the swing plane with a tremendous amount of velocity.

Left: Almost all of the player's weight has left his back foot, which is banked, and the shoulders have all but caught up to the hips.
Right: The right arm and shaft are fully extended and there's very little separation between the elbows, a sign of a powerful release.

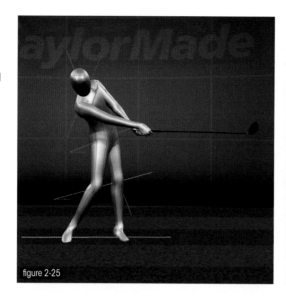

figure 2-25

figure 2-26

KEY MEASUREMENTS

Left elbow angle: 18.9°

Side spine tilt: 12.2°

Forward spine tilt: 22.4°

From a face-on view, you can still see a gap between the knees, and the left knee is almost fully extended. The shoulders are rotating on a very steep angle, which is another sign that our model golfer has done a good job of maintaining his spine angle. Almost all of the player's weight is on his left foot now, and his right heel is off the ground and banked [figure 2-26]. This could not happen if his weight was back. All of his swing's energy has been directed toward the target.

◾ POSITION 9: FINISH

You can dot the "i" with this finish: Our composite golfer looks like a pillar of balance and strength here, his upper and lower body forming

figure 2-27

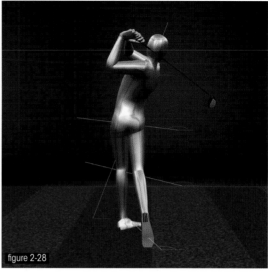

figure 2-28

one straight line [figure 2-27]. His right shoulder now points at the target (12° right of the target), and his right hip has passed his left. Both are indications of just how much shoulder and hip rotation are involved in the swing. The hips are as high as they can go, and they've moved forward just enough so that the golfer's butt is tucked underneath them. All of his weight is now over his left foot. There's only enough weight on the right foot to help with balance.

There's also a fair amount of side bend (6.3°) at the completion of the swing, but the spine has also come up some because all of the golfer's weight is forward and his shoulders and hips have leveled out. The shoulders are completely level, and the left hip is just slightly higher than the right. The knees have also leveled out and are almost touching [figure 2-28].

Left: At the completion of the swing, our golfer's spine still tilts slightly back (6.3°), away from the target. Right: Both hips are facing forward and his right shoulder points just right of the target. The thighs are nearly touching.

KEY MEASUREMENTS

Side spine tilt: 6.3°

Shoulder angle: -0.1°

Hip rotation: 110.4° open

figure 3-1

Aussie Jason Day has a picture-perfect setup.

THE DRIVER IS UNLIKE any other club in the bag, because in order to hit it effectively, you must contact the ball on the upswing. Your irons, wedges, and fairway woods/hybrids all require a descending blow, with the wedges demanding the steepest angle of attack. But the driver—because the ball sits on a tee—has to travel on a slightly ascending path in order to hit the ball flush and generate the most distance.

Our composite Tour player in Chapter 2 contacts the ball on an up path of 3°, launching the ball high with very low spin. The average golfer, however, hits down on the ball (-0.2°), despite teeing it up higher on average than most Tour pros. There are several reasons for this, but a poor setup is one of the biggest contributors. Inconsistent ball position, poor posture, and faulty alignment doom the average golfer before he even takes the club back, in many instances encouraging the very outcome (i.e., a slice) that he's most trying to avoid.

Almost everything the Tour player does in his setup is to promote the optimal upward path, from where he positions the ball and his hands to how much tilt he has with his shoulders and spine. He sets up in a very athletic posture, which puts his body in the correct angles to turn efficiently and swing the clubhead in excess of 100 mph.

In this chapter, I'll explain how you, too, can set up to have success every time you step on the tee. I say that because if there's one position the average golfer can be every bit as good as a pro at, it's the setup.

There's no reason you can't set up like Sergio Garcia or Dustin Johnson. You don't have to be super flexible or six foot four to have good posture and alignment. And you can practice the setup anywhere—at home in front of a mirror, in your living room, or on the driving range. First, though, you need to understand why the pros set up the way they do, and then you need to put these principles to work through practice and repetition.

▊ THE TWO TILTS OF THE SPINE

Throughout this book, you'll frequently come across the term "spine inclination." There have been other words used to describe this—posture, spine angle, spine tilt—but spine inclination refers to the position of the spine in its original address position to the ground. All good players, as our composite Tour model shows us in the previous chapter, maintain this inclination throughout the golf swing. There are some slight variations that occur on the downswing—as the clubhead begins to approach impact—and at the finish, but for the most part, the spine remains within a few degrees of its original inclination as the player turns back and through. This does several things: 1) It places your swing slightly behind the ball with the driver, so you can catch the ball on the upswing; 2) it creates space for your arms to swing down and for your body (hips and shoulders) to turn, which is what creates speed; and 3) it allows your body to turn in a circular motion, so that your hands and club can swing on-plane, at right angles to your spine.

At address, the spine is inclined in two directions: There's a side tilt away from the target, and a forward tilt toward the ground. The side tilt of our composite Tour golfer is 3.7°, while the forward tilt is approximately 30°. Note these "settings," since they are, no pun intended, the backbone of both your setup and your golf swing.

The side tilt does not need to be manufactured: It's a by-product of both your grip and ball position. When you grip the club, your right hand goes on underneath the left (for right-handed golfers), naturally lowering your right shoulder [figure 3-2] and putting your shoulders on

figure 3-2

a slight incline. The ball position for the driver is forward in your stance (in order to catch the ball on the upswing), opposite your left armpit. This further increases your spine tilt away from the target so that when you sole the club behind the ball, both your shoulders and spine are leaning slightly away from the target. From here, all you need to do is remain centered and you'll make contact with the ball on a slightly upward path and your head behind the ball.

The average slicer plays from a much more open stance (i.e., shoulders pointing left of the target) than the Tour player, out of fear of losing the ball to the right. This brings the spine angle more forward, toward the target, creating a negative tilt of the spine. From this position, the slicer has no choice but to increase his side tilt away from the target on the downswing in order to hit up on the ball. As a result, his weight will travel backward—not forward—on the downswing, which is a recipe for disaster. In order to hit the ball far, your momentum must be carrying you forward, toward the target; if your center of gravity is drifting backward, then the clubhead will bottom out too soon and you'll have to stand up to avoid hitting the ground a foot behind the ball. The other problem with setting up in a negative side tilt is that it can cause a shaft or hand plane that's too vertical on the backswing, with the right shoulder working out toward the target line. In this scenario, you're all but assured to come over the top of the ball on the downswing, which severely limits your clubhead speed and power.

You can have too much side tilt, too, so be careful. If your upper body is leaning too far away from the target, your shoulders will turn too flat and you'll likely hit the ball on too much of an up path. Students who set up this way usually have an up path of 6° or 7°, which can lead to a shot that spins too much and falls out of the sky quickly. It's also very hard to stay centered when your spine is leaning so far back.

The forward tilt, on the other hand, is necessitated by the height of the ball at address: At the max, even with today's high tees, the most it's going to be is 3 or 4 inches above the ground. The only way you can reach the ball and still swing your arms on a circular plane at 90° to your spine, is to tilt your upper body forward, toward the ground. This puts your shoulders on an inclined angle (i.e., steeper plane), essentially tilting this

CHECKPOINT FOR SIDE TILT

To see if you have the right amount of side tilt at address, tee up a ball and assume your normal setup. Make sure that the insides of your heels are at least shoulder-width apart, and that the ball is in line with your left armpit—or just inside your left heel. Your hands should be opposite the inside of your left thigh, and the butt end of the grip also pointing toward your left thigh. Now, without moving out of position, hang your driver straight down from the tip of your right shoulder (below). If you have the proper amount of tilt, the clubhead should fall in line with your right kneecap.

figure 3-3

circle toward the ground. If you stood straight up and moved the club in a circle, you'd swing several feet above the ball.

The average forward tilt for a Tour player is approximately 30° from vertical. This is achieved by bending forward from the hip joints, not the waist (or lower back). If you're not sure where your hip joints are, feel for the thick bone on your sides between your thigh and lower abdomen—where your upper thigh bone connects to your pelvis. This is where you want to bend from, not from above your beltline.

Besides allowing you to swing your arms at 90° to your spine, the forward tilt creates space for your body to turn and your arms to swing

CHECKPOINT FOR FORWARD TILT

Assume your stance, bending forward from your hip sockets so that your chest faces the ground. Now look down: If you have the correct amount of forward tilt, your hands should look as though they're directly beneath your chin. (This also tells you that you're standing the correct distance from the ball.) If you're looking into a full-length mirror or videotaping yourself from behind, it should also appear as though the back of your shoulder blades are in a straight line with the front of your kneecaps and balls of your feet, and the butt of the club is pointing to a spot between your belt buckle and navel [figure 3-4].

figure 3-4

A word of caution: If you look down and see your hands beneath your chin, but there's not much of a gap between your hands and body, then your pelvis is tucked in too much (under your hips). This causes your lower back and shoulders to arch into a C-posture [figure 3-5], which makes it very hard to turn behind the ball and keep your forward tilt intact. From behind, your spine should resemble its natural shape, a soft letter "s." There shouldn't be a tremendous amount of curvature to your lower back. To achieve this S-posture, feel as though you're sticking your tailbone out, as though you were trying to balance yourself against the edge of a bar stool. This will allow you to make a full-upper body turn behind the ball without losing your spine angle.

figure 3-5

down in front of you, which is how you generate clubhead speed. Tilt forward, and your hands and arms move out away from your body, freeing up space so your arms can swing down from the inside and then release to the inside, creating plenty of circular momentum. It's similar to a placekicker in football: No longer do they kick the ball straight on; rather, their leg and foot approach the ball from the inside and swing on a circular arc, producing a slight draw (right-to-left spin) and greater distance.

Your arms need room to create this circular momentum, but if you stand too erect, the hands and arms will hang too close to your body. As a result, the arms can get stuck behind you, or you could swing your hands and arms too vertically. If either happens, it's hard to recover and get the shaft back on-plane, which is how you create the most speed. When you tilt forward 30°, your arms should hang almost straight down from your shoulders, and there should be about a fist-and-a-half gap between the butt of the club and your body. Provided you have this space and you maintain your spine inclination on the downswing, you should be able to really freewheel it through impact.

BALL POSITION: PLAY IT FORWARD

Our Tour composite golfer (see Chapter 2) plays the ball directly off his left armpit, or just inside the left heel, several inches forward of center in his stance. There are two reasons why he does this: 1) so that the clubhead catches the ball on the upswing, not on the way down (which creates too much backspin); and 2) to encourage his body to move forward, toward the target. By the time the club meets the ball, the player's center of gravity (CG) is almost to his left shoulder, and the majority of his body weight is onto his left side. This is critical, because in order to hit the ball long, your CG must be moving forward; it cannot be backing up. This is no different from any other sport in which you're trying to propel an object forward, such as throwing a football or javelin, or serving a tennis ball. In each case, the object being released is going to travel farther in the air with more speed if your momentum is carrying you forward and all of your energy is directed toward the target.

figure 3-6

Generally speaking, the farther you play the ball back in your stance, the more likely you are to hang back because you have to hit up on the ball too much to catch the sweet spot. You can play the ball too far forward—many slicers do this because it points their shoulders well left of the target, away from their typical miss—but the ideal position is just off your left armpit [figure 3-6], which allows you enough time to transfer your weight forward, onto your left side, so that you can accelerate the club into a full finish position.

I should caution that when you place the ball forward in your stance, you open your shoulders to the target line. That's why it's so important to have a good side tilt in place—about 14° of shoulder tilt, 3.7° of spine tilt. If, when you turn to look at the target, you see a little bit of your left shoulder, that's good—it means you're tilting enough and your shoulders are square to slightly closed to the target.

HOW HIGH TO TEE THE BALL

figure 3-7

My oldest brother, Eddie, used to encourage me to "tee it high and let it fly." Turns out, this wasn't such good advice. In fact, on one occasion, he sliced an entire sleeve of balls into the water on the fourteenth hole at TPC Tampa, prompting us to nickname him "Eddie the Sleeve." He teed the ball so high, he had no choice but to hang back and slice it. Nowadays, he tees the ball much lower and hits it a lot straighter, but we still call him Eddie the Sleeve.

I recommend teeing the ball up so that a quarter to half of the ball sits above the top, or crown, of the clubhead (above). At this height, you're more likely to catch the ball high on the clubface's sweet spot—just above the center line on the face—with a slightly ascending blow. This is what produces the least amount of backspin and the fastest ball speeds. The higher you catch the ball on the face of today's large, 460cc heads, the less it spins and the more forward carry and roll you produce (i.e., the ball hits the ground running). Conversely, the lower on the face you make contact, the more spin you generate and the less carry and roll you produce.

You must be careful not to tee the ball too high, however, as my brother Eddie once did, because this can cause you to hang back and swing on an extreme up path. I see a lot of students with as much as 7° to 10° of up path because of the height at which they tee the ball. If you have too much up path, you most likely have too much side tilt (often the result of teeing it too high) and you'll launch the ball quickly with too much spin, which significantly reduces ball speed.

You want high launch and low spin, which you get by hitting up on the ball, but you can take that too far. Experiment with what tee height works best for you (preferably on a launch monitor) before you go making a skyscraper out of your tee and ball.

■ "PARALLEL LEFT" AND GOOD AIM

Of all the setup fundamentals, the one the average golfer struggles with most is aim. What's more troubling is that few of them realize the implication that this has in terms of generating power. Your aim not only impacts the direction your ball starts on, but also the shape your swing takes and how much curvature your shot has. It's hard to draw or fade the ball with any type of consistency if you don't have your body and clubface aimed correctly; it's also very easy to slice or mis-hit the ball if your aim is bad. Your aim is like a road map directing your body where to travel as you swing the club.

You may be surprised to learn that the average Tour player sets up to hit a slight fade. They do this because it's a much softer-landing shot than a draw, and it's easier to hold the fairways—but because they're able to keep the spin down, they can still put a good charge into the ball. The shoulders are 7° open (pointing left) to the target, and the toes, knees, and hips are virtually square to the target. The clubface also points at the target. The composite golfer's shoulders and the rest of his body are aligned "parallel left" of the target line, some 20 to 30 yards left of the target.

The average over-the-top slicer sets up with their toes pointing right of the target and their shoulders aiming well left of the target.

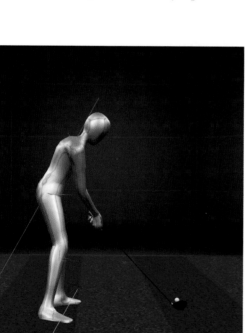

figure 3-8

Your weakest hitters (i.e., slicers), set up with their shoulders and hips pointing farther left—to avoid hitting the ball to the right—but they aim their feet directly at the target, not parallel left of it [figure 3-8]. They also point the face away from the target, in a slightly closed position, because that's where they want the ball to start (i.e., left of the target line). This is the prototypical setup for your over-the-top slicer: The hips and shoulders are well

CHECKPOINTS FOR AIM

One of the best ways to check your aim is to set up in front of a full-length mirror. First, assume your setup and look down to see that your toe line is parallel left to your clubface, which should point at an imaginary target. As you look back into the mirror, you should see the following: 1) knees even and over the balls of the feet, parallel to the target line; 2) forearms even, parallel to the knees, hips, and target line; and 3) shoulders even, with the shoulder blades in line with the front of the kneecaps and the balls of the feet (below). It's okay if you see a trace of your left forearm, which would indicate that your shoulders are slightly closed. What you don't want to see is the underside of your left forearm below your right, because that would mean that your shoulders are significantly open.

figure 3-9

open to the target line, but the feet are pointing directly at the target. From this position, the golfer has little option but to come over the top of the ball (i.e., swing from out to in, across his body), because if he swings down to the ball from the inside the ball will sail off the planet to the right (because his feet are aimed right of the target). Unfortunately, the ball winds up peeling off to the right anyway, because the face is open relative to the club's steep, out-to-in path, and this causes the ball to start left of the target line and then curve sharply to the right. He also can't generate any clubhead speed because he's pulling the club down across his body, instead of swinging freely out toward the target.

The average golfer has no concept of what parallel left is. He doesn't understand that he's standing several feet to the side of the ball, and that the target should appear as though it's well to the right of where his feet are aimed. Instead, he aims his feet at the target as though he were standing directly behind the ball, not knowing that what he's doing is aiming to the right and putting himself in position to swing from out to in.

When you assume your stance, the first thing you should do is aim your feet at a spot 20 to 30 yards left of your target. Find a tree, the edge of a bunker, or something to point your toes at. Now imagine a line drawn from your toes extending all the way out to this spot: This line should appear as though it's running "parallel left" of your target line, like the inside rail on a train track. It should never converge with the other line (if it does, you're aiming your feet to the right). Next, point the clubface at your target and align your shoulders, forearms, and hips on the same line as your toes. If your shoulders look closed to you, that's okay—the more they point to the right, the more likely you are to swing the club down on the proper inside path. Your alignment should encourage an inside path, and provided that you aim your feet and shoulders correctly—parallel left with the feet square to slightly open at the shoulders—you should be able to swing the club down on-plane with maximum clubhead speed.

◼ CLUBFACE AIM AND ITS IMPACT

The position of the clubface at address also has a great influence on the path the club takes during the swing. Almost all slicers address the ball with a closed clubface, because they don't want it to go right. They want the ball to start left of the target, in the opposite direction of their miss. In closing the face, the handle gets pulled away from the target, promoting an outside takeaway and the lifting of the club, and an out-to-in downswing path. If the clubface is set open, the handle is moved toward the target, encouraging a shallower, more on-plane takeaway and an in-to-out swing. Faders of the ball tend to set the face slightly closed; those who draw the ball like it slightly open.

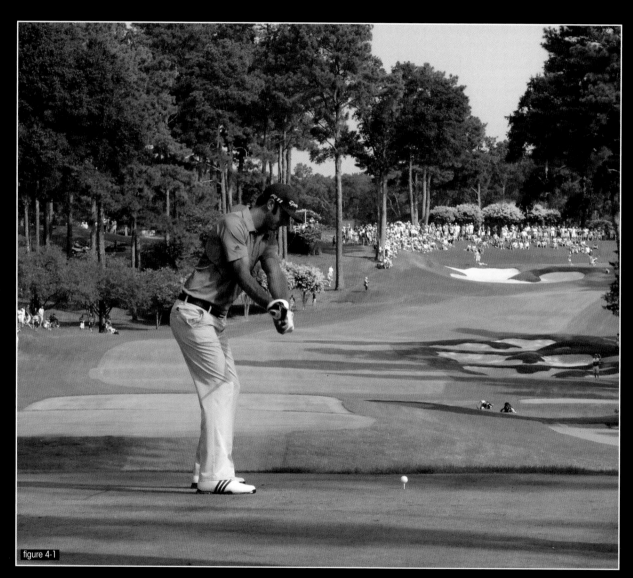

figure 4-1

A flat left wrist, which Dustin Johnson has here, is the key to an on-plane takeaway.

CHAPTER 4

The First Few Feet of the Swing

THE FIRST HALF OF the backswing is similar to the first few feet of a 100-meter race: If you get off the blocks too slowly, or you do something to alter your running style, you'll have a hard time recovering and generating the type of speed you need to win the race. The golf swing is not a race, but the first few feet helps set the stage for what comes after, especially at impact. Everything you do in the swing is geared toward putting you in the best impact position possible, and if you cup your left wrist right out of the gate or come up out of your spine angle, then you have to recover somewhere to get back into the correct position. That's not an easy thing to do in the 1.25 seconds it takes to reach impact. The fewer compensations you have to make in the swing, the more efficient and powerful your swing will be.

What you do during the first part of the backswing also affects how much width, coil, and power you're able to produce later on. The average Tour player is able to make a full backswing turn and store momentum for the downswing without losing his original spine inclination. There are several things that have to take place early on for him to do this effectively, and I'll get into each of those in this chapter. I'll also look at some

figure 4-2

of the key mistakes amateurs make during the takeaway, such as swinging back with their hands, and explain how they can get these things corrected so that they, too, can achieve a powerful impact position.

FIRST MOVE BACK: LEFT SHOULDER GOES DOWN

One of the most frequent questions I get from golfers is, "What starts the swing?" In studying the composite Tour swing avatar and the motions of hundreds of Tour pros, it's fair to say it's not one independent action of the hands, arms, or shoulders, but a one-piece movement of the hands, arms, and shoulders away from the target. The triangle created by the arms and shoulders at address moves in unison for the first foot or so, with very little movement from the hips. As this happens, the left shoulder tilts down toward the ground and the left arm swings across the chest [figure 4-2]. As a matter of fact, the whole left side, including the left hip and knee, starts to drop down as the clubhead swings away from the ball. Consider that from address to mid-back (i.e., shaft parallel to ground), the left shoulder angle drops 38.2° while the left hip dips 6.6° and the knee 3.9°. While this happens, the right side extends upward in the opposite direction.

The left side shrinks for a reason: It's the only way you can maintain your spine inclination (both forward and side tilts) and still make a full shoulder turn. If it didn't, you'd have to rise up out of your spine angle, leveling out the shoulders. Ninety-nine percent of all Tour players maintain their spine inclination as they turn behind the ball. It's how they swing their hands and shaft on-plane and create coil and momentum for the downswing, which is key if you want to generate maximum clubhead speed and power.

3 MUSTS FOR A GOOD TAKEAWAY

If you look at the avatar of Sean O'Hair in the mid-back position from both straight on and down the line, several things should stand out:

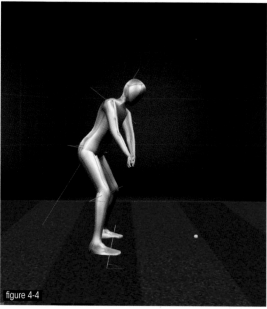

figure 4-3

figure 4-4

1) both arms are nearly extended [figure 4-3]; 2) the back of his left wrist is flat as a pancake; and 3) his original spine inclination remains virtually unchanged [figure 4-4]. This doesn't happen by accident. Tour players spend a great deal of time on their takeaway, because they know how important it is to get the swing off to a good start—i.e., with the club on-plane, the face square, and with good width. Here's a further look at these three key backswing moves, and how they equate to more power.

■ RIGHT ARM STRAIGHT

For the first several feet of Sean's swing, he's able to keep his right arm relatively straight. There's a little bend to the elbow (just 20° in the mid-back position) and the triangle formed by his arms and shoulders at address is still intact, as is the gap between his forearms and elbows. This is a great move to copy, because if you can keep the right arm relatively straight for the first few feet of the swing, the left arm will also extend, creating tremendous width to the backswing [figure 4-5]. Width is a

figure 4-5

measurement of the distance between the butt end of the grip and your chest, and the wider this gap, the farther the clubhead is from your body and the more speed you'll be able to produce. The clubhead looks miles away from Sean's body when the shaft is parallel to the ground: That's a move—provided you maintain your spine inclination and don't move your head off the ball—that will bring you a lot of power.

Most amateurs pick the right arm up at the start of the backswing, forcing the right elbow to bend and the shaft to move off-plane. This shrinks the gap between the clubhead and your body and opens the face, two moves that hinder your ability to hit the ball far. The average golfer's right elbow angle is 41° in the mid-back position, twice as much as O'Hair's. This grows to 76.1° in the nine o'clock position (left arm parallel). That's why, when viewed straight on, the average player's hands are much

figure 4-6

The average slicer folds the right arm (41°) too much in the mid-back position, which narrows the gap between the hands and body and robs him of distance.

closer to their body than the Tour player's hands throughout the first half of the backswing [figure 4-6]. Consequently, the arc that the club-head travels on is much narrower than the Tour player's, diminishing the amount of speed and distance he can produce.

DRILL: SQUEEZE THE FOREARMS TOGETHER

To get a feel for keeping the right arm straight on the takeaway, lodge a ball, pillow, or rolled-up towel (pictured) between your forearms and assume your natural setup position [figure 4-7]. Swing the clubhead back until the shaft is parallel to the ground, maintaining the triangle formed by your arms and shoulders at address. As you swing back, squeeze the towel by pushing your forearms and elbows together [figure 4-8]. This will prevent the right arm from folding, keeping the arm extended. On the course, feel like you're applying pressure with your forearms to this imaginary towel, and you'll widen your swing arc and boost your power.

For a demonstration of this drill and all of the other drills in this book, as well as the checkpoints, please visit www.3dgolfer.net.

figure 4-7

figure 4-8

■ LEFT WRIST FLAT

The back of O'Hair's left wrist in the mid-back position is slightly bowed out, in the opposite direction that many golfers move their wrist. They cup the wrist almost immediately on the takeaway, creating an angle between the back of their hand and the left forearm. This is a direct result of starting the club back with the hands rather than in a one-piece movement with the body, as most Tour players do. When the hands take over, the forearms rotate too much, opening the clubface and pulling the clubhead to the inside and off-plane [figure 4-9]. From here, it's very easy to dump the club (i.e., release it early) on the downswing, because you have to close the face for it to be square at impact. That's why cupping the left wrist at the start of the swing often leads to a chicken-wing finish (bent left elbow), because the early release forces the arms to swing down across the body.

figure 4-9

The following drill will teach you how to keep your left wrist flat on the takeaway, thus preventing you from cupping the wrist and rolling the face open. Slip a ruler or pen under the wristband of your watch so it lays flat against the back of your left wrist and forearm. Take your address position and swing the club back slowly until the shaft is parallel to the ground (right). The pen should remain comfortably flat against your wrist. However, if you cup the wrist, you'll feel some pressure on the back of your hand. (If you're using a pencil, be careful as the tip could puncture your skin.) Repeat several times, then remove the pen and hit a few drives, maintaining the feeling of a flat left wrist during the initial stages of the backswing.

figure4-10

All good ballstrikers keep the back of the left wrist flat in the early stages of the backswing. The wrist is so flat, in fact, that you can lay a square divot on the back of the left forearm and wrist. The left forearm rotates slightly, but only in response to the chest and shoulders turning away from the target. As a result, the shaft is on-plane (the clubhead is in line with the hands, not inside the hands) and the clubface is square, matching the angle of the spine. If you can achieve this position, you'll have a much better chance of delivering the clubface square to the ball at impact. The straight left wrist also helps you keep the extension in your right arm and maintain the triangle between your arms and shoulders, creating more width and allowing you to swing in a circle. If the left wrist starts to cup, the arms lift up and the body doesn't turn as effectively. There's too much of an "up" motion to the swing and not enough "around."

figure 4-11

SPINE ANGLE STAYS INTACT

Sergio Garcia maintains his spine angle during the early part of the swing, which promotes a wide takeaway and more power.

I touched on this earlier, but I cannot stress enough the importance of maintaining your spine inclination throughout the entire swing. It's not an easy thing to do, but the average Tour player is able to rotate his body and extend the clubhead far away from his chest without significantly changing his forward or side spine tilts [figure 4-11]. As a result, he's able to turn his shoulders deep behind the ball, with a square clubface, setting the stage for a powerful, on-plane downswing.

The most common early backswing fault I see with amateurs is that they stand up out of their posture. The hands work up, the arms pull off the chest and the spine straightens, which levels out the shoulders [figure 4-12]. From here, to hit the ball solidly, you have to dip down and restore

figure 4-12

DRILL: HEAD AGAINST WALL

To train your body to remain in its original spine inclination as you swing back, try the following drill, which you can do at home or on the course with the help of a golf cart. Stand facing a wall, with your toes approximately a foot from the wall. Assume your address position, tilting forward from your hips until your forehead rests against the wall. Pretend you have a club in your hands and swing your arms back until your left arm is parallel to the ground [figure 4-13]. Keep the back of your left wrist flat, and the distance between your elbows small. Since your head is propped up against the wall, it won't be able to move backward or up; the left shoulder will automatically turn down and you'll remain in your original spine angle [figure 4-14].

figure 4-13

figure 4-14

your original forward tilt, which forces the hands out away from the body on a steep out-to-in downswing path. There are a number of things that can cause you to stand up, but the biggest culprits are a poor setup and a desire to turn more than your flexibility will allow. It's much easier

figure 4-15

to turn your shoulders if you stand up and your spine gets vertical, and most golfers will do anything to get more shoulder turn.

WRISTY BUSINESS

Most Tour players try not to set their wrists too early on the takeaway, because they're trying to establish as much width to their backswing as they can. Width creates speed, because the farther you extend the club-head away from your body—without moving your spine—the wider the swing's arc and the more distance the clubhead has to travel. The average golfer, however, lifts the club up with his hands instead of keeping the left wrist flat and the shaft on-plane. As a result, the wrists set early and the hands and club move on a plane that's too vertical, narrowing the swing's arc [figure 4-15].

The bending of the right elbow is one of the biggest power leaks in the swing. The wrists should hinge naturally as the club swings upward, not as a result of the elbow folding.

The cocking angle, which is a measurement of the angle formed by the clubshaft and the left forearm, is virtually the same for the average player in the mid-back (39.2°) and shaft parallel (80.2°) positions as it is for the Tour player (44.2° and 80.1°, respectively). But how the average golfer goes about achieving this angle is entirely different. He does it by folding his right elbow. This increases the cocking angle, creating the appearance that he's hinging his wrists properly. However, he won't be able to maintain the angle on the downswing for very long, because the clubface is open and he'll have to release it early to get it back to square.

CHECKPOINT FOR WRIST HINGE, WIDTH

To see if you have the right amount of wrist hinge and width in the mid-back position, swing the shaft back to parallel, keeping your left wrist flat and your head still. Look down: You should see your left hand to the outside of your right little toe (below). If you pick the club up with your hands, the left hand will appear well inside the right foot, and if you move your head off the ball (i.e., sway), then the hand will appear well outside your right foot. From a straight-on view with the shaft parallel, the left arm should be straight and the butt of the club just over the outside of your right foot (please refer back to figure 4-2).

Many golfers, in an attempt to get more width on the backswing, move their upper body too far off the ball. When this happens, the shoulders become level and it's almost impossible to get back to your left side fast enough to make solid contact. It's very important to generate width in your backswing, because you'll hit the ball farther, but only if you remain in your original spine inclination.

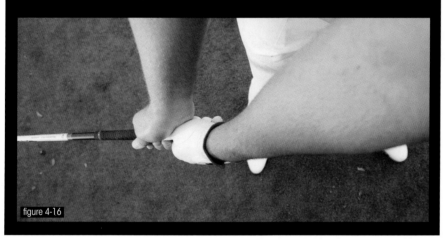

figure 4-16

Wrist hinge isn't something you create, nor is it something you should think about. Provided you keep the left wrist flat and the shaft on-plane, the wrists will hinge naturally, on their own, in response to the club swinging upward. Focus on creating width early on your backswing, and the hinging will take care of itself.

figure 5-1

As the club reaches the top of his swing, Sergio Garcia's weight is already moving onto his front side.

TOP OF THE BACKSWING

CHAPTER 5

Turn, But Don't Over-Rotate

IF THERE'S ONE POSITION that golfers view as a precursor to how much power they can produce, it's the top of the backswing. Most golfers believe that if they turn their shoulders 90° or more, good things will happen. When I ask a student what they're trying to do on their backswing, the response I typically get is, "turn my shoulders more." People confuse turning with power; they think they can't turn their shoulders enough, and so they try to turn them more and more. What they don't understand is that there are rules that apply to this line of thinking, just as there are restrictions on that too-good-to-be-true $89 airfare they saw publicized online.

The reality is that the average golfer turns their shoulders enough. In fact, would you believe 117°? That's not a misprint, but the average of several hundred 15-plus-handicappers (collected on the MAT-T system) who slice the ball. They also turn their hips 56°, which is 7 more degrees than the average Tour player does. So why do their tee shots travel about a football field shorter than the average Tour player's? The answer has to do with those rules. There are things the Tour player does to maximize power, and things the average player does to take it away. In this chapter,

I'll discuss all of these factors, and explain how you can plug the power leaks on your backswing so that you, too, can add 15 to 20 yards to your drives and reach more par 5s in two.

THE TRUTH ABOUT THE X-FACTOR

The X-Factor is a term invented by noted golf instructor Jim McLean that describes the differential between shoulder turn and hip turn. Research conducted by McLean found that all great ballstrikers turn their shoulders more than their hips at the top of the backswing [figure 5-2]. The bigger the X-Factor, the more clubhead speed your body will be able to generate and the farther you'll hit the ball.

figure 5-2

This is all true, but *only* if you maintain your original spine inclination from address [figure 5-3]. It does you no good to turn your shoulders 100° and your hips 40° if you come up out of your spine angle on the backswing. If you stand up, you're not going to rotate your body through impact like you would if you stayed in your spine inclination, nor will you be able to keep the club on-plane. You're going to have to restore your spine angle in order to get down to the ball, which will cause you to dump the clubhead (also known as "casting" or an "early release") and come over the top of the ball on the downswing. This casting action forces the hips and shoulders to rotate too soon, long before impact, which slows the clubhead down. It's like a sprinter pulling up lame with a hamstring injury 60 meters into a 100-meter dash. You need to have the hips and shoulders rotating through impact to create speed.

figure 5-3

The goal of the average golfer should be to turn their shoulders as far as they can to the top while maintaining their spine angle [figure 5-3]. That's more important than having a giant X-Factor. If all your flexibility will allow you before you lose your spine angle is 25° of hip turn and 70° of shoulder turn, then that's your most efficient backswing. You'll hit the ball farther than you would if you came out of your spine inclination and turned your hips 55° and shoulders 115°. You'll make more solid contact, too.

There's nothing wrong with McLean's theory, it's just that people over the years have misinterpreted the X-Factor to mean "how much" the shoulders should turn. They think a big shoulder turn guarantees more power, and so they do whatever they can to maximize their turn, even if their flexibility won't allow for it. As a result, they over-rotate their hips and shoulders. While they might think this move is powerful, it's just the opposite. Not only does it throw off their downswing sequencing, but it severely limits the amount of coil, width, and power they can build with their backswing. Very flexible people can also turn too much, making it very difficult to not increase their spine angle on the backswing. This is a common problem with women, who tend to over-rotate their shoulders and hips.

As long as you maintain your spine angle, your shoulders will still out-turn your hips and you'll have your X-Factor. It might not be as big as you'd like, but it will put the club in a better position to swing down to the ball on the correct path—from the inside. It will also help you store more torque and energy so that your hips and shoulders can unwind naturally through impact, providing an additional burst of speed to the club. Our average Tour composite golfer has 130° of shoulder turn and 49° of hip turn, nearly a 3:1 ratio, yet his forward tilt increases only 1° and his side tilt 0.5°. In other words, he's able to maintain his spine inclination despite wielding such a huge X-Factor. So how is he

Dustin Johnson, one of the longest hitters on the PGA Tour, turns his shoulders a whopping 144° to the top while actually dropping his spine angle down toward the ground. Most golfers have to stand up to turn their shoulders more than 100°.

figure 5-4

able to do it? A glimpse at Dustin Johnson's backswing avatar holds the key. Johnson, who is one of the three longest hitters on the PGA Tour, turns his shoulders an amazing 144° and his hips just 43°—as big an X-Factor as there is—and he does it without losing his forward tilt [figure 5-4]. Actually, Johnson's spine angle goes *down* 4.3°. This speaks to Johnson's freakish flexibility—I don't know of any Tour player who's able to increase this angle down so much and still turn his shoulders as far as he does—and it provides a blueprint for how to make a proper turn.

While I don't recommend taking it to the extreme that Johnson does, you're much better off dropping your spine angle down toward the ground than you are picking it up. Most golfers do the latter to accommodate a big X-Factor; they stand up and lift their arms to turn their shoulders. This pulls the left shoulder up and levels out both shoulders, which destroys your width and coil and leaves you in a very weak position to swing down to the ball [figure 5-5].

figure 5-5

To maintain your spine inclination, the left shoulder must move down on the backswing. It can't go up. The left hip and knee must also drop down—if they didn't, there's no way you'd be able to turn your shoulders and still maintain your forward tilt to the ground. The average Tour player's left shoulder rotates more than 120° from its original address position to the top of the swing. At address, it's almost in line with

CHECKPOINT: LEFT ARM HIDES SHOULDERS

When viewed from behind, your left arm should be on the same line as your shoulders—virtually hiding your shoulders—at the top of the swing [figure 5-6]. This is significant because if your left arm is on-plane, then you more than likely maintained your forward tilt toward the ground, with no upward movement of the spine. Your hands should be deep, over your right biceps, which allows you to swing your hands and arms down on a straight line to the ball from the inside, without having to change your spine angle or reroute your hands. The purpose of the backswing is not only to store power, but to get the club in a good position to swing down to the ball on the correct path, 90° to the spine. If you've got these two checkpoints in place, then you should be able to do this consistently enough, and you'll not only hit more fairways, you'll drive the ball farther, too.

If you look at the avatar of our typical over-the-top slicer, you'll see what happens when you come up out of your spine angle [figure 5-7]. Not only does the left shoulder come up, but the hand plane does as well, pulling the left arm off your shoulder plane. The club is so vertical now that the hands are closer to the right ear than they are the shoulder. From here, you're in no position to swing the club down from the inside. Unless you pull a Jim Furyk and loop your hands back to the inside on the correct path, the club will swing out toward the target line and cause you to come over the top. But if you get the hands in the right spot, with the left arm mirroring the angle of the shoulders, you'll be ready to let it fly on the downswing.

figure 5-6

figure 5-7

Justin Rose's left arm remains on his shoulder plane [figure 5-6], while the average over-the-top slicer lifts the arm off-plane [figure 5-7].

the left ear, but at the completion of the backswing it's directly under the chin and over the inside of the right leg. It turns down and across his chest, and points at the ground (outside the right foot) at the top of the swing. The left hip drops 20° and the left knee nearly 11°. The whole left side shrinks to accommodate the spine.

Therefore, if you want to hit the ball farther, don't think about turning your shoulders as much as keeping your left shoulder down and your butt out—you don't want your pelvis moving in toward the ball. Make swings without a club with your head up against a wall, turning your left shoulder down and across your stance line. If your head moves off the wall before you reach the top of your imaginary swing, then you don't have enough flexibility to make a full shoulder turn and maintain your spine angle. That's okay: See how far you can rotate your shoulders before your head leaves the wall. This will tell you just how far you should turn your shoulders on the backswing. Most of you won't like this initially because your backswing will be much shorter than you had hoped. But don't try to turn them any farther, because this will force you to raise out of your spine angle, greatly reducing your power. Provided that you stayed in your address posture, you should still feel a good stretch between your shoulders and hips and a tight coil, which will put you in a strong position to create speed on the downswing.

ANGLING FOR MORE POWER

In the previous chapter, I introduced you to the term "cocking angle," which is a measurement of the angle formed by your left forearm and the clubshaft. As the club swings upward, the wrists begin to hinge naturally, increasing the size of this angle. The wrists should stop hinging when the left arm reaches parallel on the backswing, but the cocking angle will continue to grow due to a subtle movement at the top of the swing. This move is hard to see, but it has a big influence on how much clubhead speed you're capable of producing on the downswing.

If you look at the two avatars of Sergio Garcia side by side [figures 5-8 and 5-9], you'll see a jump in the cocking angle of almost 30° from the nine o'clock position, or left arm parallel (73.4°), to the top of the backswing (102.7°). Garcia does this by shifting his center of gravity forward, toward his left foot, as he's completing his backswing. While his hands are moving up ever so slightly, his lower body is inching

figure 5-8

figure 5-9

As Sergio Garcia completes his backswing, he shifts his body's center of gravity forward, toward the target. This increases the angle between his left forearm and club shaft by almost 30°, creating more lag and power for the downswing [figures 5-8 and 5-9].

toward the target, increasing the angle while also expanding the lag (i.e., the distance between the clubhead and the hands).

By the time Sergio starts his downswing, almost all of his weight is on his left foot. Why is this so important? Because with his weight on the left side, he should be able to hold the cocking angle much longer on the downswing, which will allow him to release all of that stored up energy when the wrists unhinge and the angle straightens out through impact. The average golfer is prone to lose his spine angle at the top and begin the transition with his center of gravity still over his back foot. As a result, he has a very difficult time getting back to the ball (i.e., to his left foot) by impact. With most of his weight still behind the ball as he starts down, it's nearly impossible for him to hold the lag. The arms take over, dumping the angle well before the clubhead reaches the ball.

To hit the ball far, you need the clubhead to be accelerating when it meets the ball. Dumping the angle prematurely—before the clubhead gets to the ball—shortens the swing arc, thus slowing the head down; holding the angle deep into the downswing and then releasing it through impact speeds it up.

LEFT HEEL STAYS DOWN

Somewhere along the line, you might have read something to the effect that, if you have limited flexibility, it's okay to lift your left heel off the ground on the backswing, à la Jack Nicklaus. This is true, provided you're able to stay in your forward and side tilts and the left knee goes down. Lifting the heel will allow you to turn your hips and shoulders a little more, so you can generate more clubhead speed and power on the downswing.

In my experience, however, people who lift their left heel off the ground also can shift too much weight to their right side and keep it there too long. This limits their ability to transfer their weight forward and causes them to swing over the top with very little clubhead speed or power. Golfers who lift their left heel also tend to collapse the left knee and lose their spine angle—usually in the form of a reverse pivot (i.e., the spine tilts toward the target at the top of the backswing). From here, the average golfer struggles to restore his spine angle and, thus, hit the ball solidly. If you look at the avatar of Sergio at the top of his swing [figure 5-9], you can see that his left knee moves in ever so slightly, so that it's pointing just in front of the ball. The knee moves mostly down, instead of buckling in toward the right foot, and the left heel remains on the ground. Furthermore, the space between his knees doesn't change from his address position. That's a great image to carry with you: If you want to maintain your spine angle and create maximum width and coil on your backswing, then keep your left heel planted and your knees far apart. If you want to do something to facilitate a bigger, deeper turn, drop your right foot back slightly and flare the foot open about 45°, which will preset your shoulders in a closed position and clear your back hip out of the way so you can rotate both a little easier.

figure 6-1

Few players exhibit more clubhead lag at the start of the downswing than Sergio Garcia.

THE FIRST MOVE DOWN

OKAY, SO YOUR BACKSWING is textbook: You've made a full shoulder turn, maintained your spine angle, and kept your hands and clubshaft on-plane. You're home free, right? Not so fast. You can do everything perfectly on the backswing, but if you don't get your sequencing just right at the start of the downswing, you can still come over the top of the ball and hit a weak drive. A good backswing puts you in the best position to attack the ball from the inside, but it doesn't guarantee power. There's a lot that can go wrong in the quarter of a second it takes for the clubhead to reach the ball.

There is a sequence to the way your hands, arms, hips, and shoulders move at the start of the downswing; however, this sequence is counterintuitive to the way most people throw a ball or hit an object, because it requires you to use your smaller muscles, not your shoulders. What's more, it's easy to get overanxious at this juncture because you want to hit the ball and see what the outcome is. But the faster you start down, the harder it is to keep the club on-plane and create speed. To hit the ball far, the clubhead has to gradually build speed on the downswing, reaching its maximum velocity as it reaches impact. It can't do that if your sequencing and timing are off.

In this chapter, I'll take a closer look at the downswing sequencing of the Tour player versus the typical over-the-top slicer, and also reveal how the longest hitters are able to increase their lag and create even more clubhead speed through impact.

▊ HOLD THE SHOULDERS

The impact position is often referred to as the "moment of truth," but for a lot of golfers, the pivotal moment occurs a fraction of a second earlier, at the start of the downswing. What you do here will have a huge influence on the path the club takes into the ball, as well as how much clubhead speed you're able to generate.

Your shoulders hold the key. From the top, your instincts tell you to swing the club down with your shoulders, because they feel very powerful and are naturally the first to move anytime you want to hit something hard. But in the golf swing, the moment your right shoulder moves out toward the target line and your shoulders open up, you're dead. This move throws the clubhead over the top, which forces the right arm to extend too soon and the club to cut across the ball from out to in. There are other reasons why you come over the top, which I'll get into shortly, but if you want to maximize your power you have to keep your shoulders passive during the early part of the downswing.

So how do you do this? First of all, you need to get your weight moving forward; then your hands will fall on-plane (i.e., directly over your right biceps) at the start of the downswing. From here, they can take a straight line into the ball, which is the fastest, most effective way to swing the club down—not with your shoulders. Your hands and wrists are the primary speed producers in the swing, and because they have more distance to travel, it makes sense to get them moving first from the top of the swing.

Once you arrive at the top of your swing, focus on moving your weight forward. Keep your back facing the target and accelerate the handle in a straight line toward the ball, as though you were hammering a nail with the butt end of the club. Your hips should respond to your center of gravity moving forward and the swinging of your arms and begin to unwind, at which point your shoulders should also start to ro-

figure 6-2

figure 6-3

tate. If your shoulders are the first to unwind, your hands will swing out in front of your right shoulder, pulling your right elbow off your rib cage and pushing the handle out toward the target line [figure 6-2]. From this position, it's virtually impossible to swing the club down from the inside, not to mention rotate your hips and shoulders through impact.

If you look at the image of Justin Rose above, you'll see that his hands remain in front of his right biceps during the initial transition phase of the downswing [figure 6-3]. His back is still turned toward the target, and his shoulders are closed to the target line. This is nearly a mirror image of his backswing in the same position, with the left arm parallel to the ground. His right elbow points down and the butt end of the club also points at the ball. The image here of Rose is in stark contrast to our over-the-top slicer, whose hands are well in front of his right shoulder now, directly under his right ear. Viewed side by side, you can see just how much closer the slicer's handle is to the target line than Rose's, and how much steeper his downswing has to be.

DRILL: RIGHT ARM TO CHEST

The minute your right shoulder starts to come over the top, your right elbow backs away from your body and your upper arm loses contact with the side of your chest. But if you start down correctly—moving the handle of the club straight at the ball—your right arm maintains pressure with your chest, keeping the hands on-plane and the shaft on the proper inside path [figure 6-4].

As a drill, lodge a head cover between your right biceps and the side of your rib cage, and assume your normal stance. Swing the club back to the top and then down, maintaining pressure between your arm and the side of your chest. If your sequencing is good at the start of the downswing, the head cover should stay in place [figure 6-5]. If you start down with the shoulders, however, your right elbow will pull away and the head cover will fall out. On the course, feel free to tuck your shirt in under your upper arm as a reminder to keep pressure against your rib cage as you start down. As long as this pressure is there, your hands should remain on-plane and you should be in good position to hit the ball with some authority.

figure 6-4

As you start down from the top, your upper right arm should maintain some pressure against your body, as Sergio Garcia demonstrates here (top). If your right shoulder moves out toward the target line, you'll come over the top.

figure 6-5

THE SWING'S DEATH MOVE: COMING OVER THE TOP

There are a number of factors that can cause one to come over the top, which is the most crippling mistake that amateurs make at the start of the downswing. The instinct to throw the shoulders, which I mentioned earlier, is one of them, but the three biggest culprits I see are a poor weight transfer, a cupped left wrist (on the takeaway), and a loss of spine angle.

The average slicer can't get their weight to their front foot fast enough during the transition. There's too much weight on the back foot to start with, which causes their hands and the shaft to kick out and their shoulders to open too quickly. Nothing good can happen on the downswing unless you first get your weight to your left side, because the clubhead will more than likely bottom out too soon, behind the ball, and you won't be able to control the face angle at impact.

Most weak hitters typically cup their left wrist on the takeaway, too, snatching the club up with their hands. This action causes the forearms to roll and the face to open, making it difficult to swing the club on-plane and return the face to square at impact. At some point, you have to close the face, and the only way to do that is to release (unhinge) the right wrist early on the downswing, which causes the right elbow to straighten too soon and the clubhead to come over the top. Not only does the club move off-plane, but the angle created by your left forearm and the shaft gets dumped, tossing away all the additional leverage and speed you had built up on the backswing.

I touched on the ramifications of a poor spine angle in the previous chapter, but when you come up out of your original spine inclination on the backswing (often the result of trying to turn the shoulders too much), you have to restore this angle to the ground on the downswing. If you don't, you'll miss the ball entirely or make a glancing blow. In restoring the angle, the body has to come forward, toward the target line, which throws the club over the top.

DRILL: BUMP THE SHAFT

What's the best way to not swing over the top? First and foremost, you simply must get your weight to your front foot first before swinging your hands and arms down. You have no chance if you start the downswing with your weight on your back foot. Here's a drill that will get you to feel where your weight needs to be during the transition, so that your arms and the shaft can swing down on-plane instead of coming over the top.

Stick a shaft in the ground so that it's about 3 inches outside your left hip when you take your setup. Swing the clubhead back to the top [figure 6-6], bumping your left hip forward just enough to touch the shaft as you're completing your backswing [figure 6-7]. Hold this position for a second, and then start down. As you change direction, try not to move your hands too much. Repeat several times, getting a feel for how your weight moves into your left side before you start the downswing. This forward shift in momentum increases the angle between your left forearm and the shaft, and also adds lag—i.e., more distance between the clubhead and your hands—making it easier for you to hold this angle well into the downswing.

figure 6-6

figure 6-7

GRIP CHANGER

If you've worn a hole out in the palm of your glove near the thumb pad, you could be having regripping issues. Many golfers, in an attempt to create more power, unhinge their wrists and let their fingers separate from their palms as they near the top of their backswing. This forces them to regrip the club on the transition—i.e., grab it tighter—which can lead to an early dumping, or release, of the clubhead. It also can prevent you from shifting your weight forward, which causes you to stay on your back foot and swing the club off-plane. Once the club moves off-plane, it's hard to control the face and generate any clubhead speed.

Whatever your grip is—neutral, strong, weak, ten-finger—it's important not to hold the club too tight. On a scale of 1–10, your grip pressure should be about a 3 or a 4—just enough tension to keep the grip from turning in your hands. If it's significantly higher—think death grip—your arms will be too rigid and you'll have difficulty rotating your forearms and squaring the clubface at impact.

▉ HOW TO INCREASE YOUR LAG

In the previous chapter, you learned that the Tour player shifts his center of gravity forward, onto his left side, as he completes his backswing. This subtle movement of weight at the top of the backswing increases the distance, or lag, between the clubhead and his hands, making it easier to retain the angle between the shaft and the left forearm for a longer period of time on the downswing. Maintaining this angle (known as the "cocking angle") is extremely important, because as the angle is released (i.e., straightens out) through impact, it sends an additional burst of speed and energy to the clubhead. If the angle is released too soon, which happens when you come over the top, you lose out on all of this potential speed and power.

As he starts down from the top, the average Tour player actually increases this angle a few degrees more. The reason for this is that he's still shifting his center of gravity forward, toward the target, while his hands and arms are swinging down in the opposite direction. His body is being pulled in two completely different directions, thus stretching the lag out even further. Take a look at the avatars of Sergio Garcia on the next page [figures 6-8 and 6-9]: At the start of the transition, his cocking angle is 102.7°, but by the time his left arm reaches parallel on the down-

figure 6-8

figure 6-9

At the start of the downswing, the clubhead lags far behind Sergio Garcia's hands [figures 6-8 and 6-9, above]. The angle between his left forearm and clubshaft actually increases before eventually straightening out through impact, sending an additional burst of speed and energy through the shaft to the clubhead [figure 6-10, opposite page]. The average slicer hangs back and releases the angle between the shaft and left arm too soon, causing the clubhead to lose significant speed before it gets to the ball [figure 6-11, opposite page].

swing, it's 108.7°. He's increased the angle by 6°, whereas the average over-the-top slicer usually loses 5°. Just prior to impact, when the shaft is parallel to the ground on the downswing, Sergio still has 70° of cocking angle [figure 6-10], whereas the slicer (62.5°) is fighting to keep the angle [figure 6-11]. The slicer's about to lose the lag because he's hanging back; he has yet to shift his weight forward, which will cause him to dump the angle early and lose distance.

To sustain this lag longer, you have to move your center of gravity forward—to the front foot—while remaining in your original spine inclination at the start of the downswing. With most over-the-top slicers, their weight never leaves their back foot. Their spine angle straightens and they're forced to extend their right arm quickly in an attempt to square the clubface and create power. That's why their cocking angle is less than the Tour pro's just prior to impact. Work on the previous "Bump the Shaft Drill" (see page 70) and "Ball Under the Bridge Drill" (see Chapter 7) to help you with your weight transfer to your left side. Then practice the following two "Wide-to-Narrow" drills to increase the lag in your swing. The bigger the lag, the more clubhead speed you'll generate at the time you need it most—impact.

figure 6-10

figure 6-11

WIDE-TO-NARROW DRILL I

If you were to trace the arc the clubhead takes during the swing, it would form the shape of a half-moon, going from very wide on the backswing to narrow on the downswing. The narrow downswing track is the product of a good transition: The hands move down in a straight line toward the ball, which brings the handle in closer to the body and tightens the arc. This, in turn, increases the lag between the clubhead and the player's hands, storing more power to be released through the ball. The average golfer throws the clubhead over the top (above the correct plane) at the start of the downswing, essentially retracing the wide path it took on the backswing.

figure 6-12

To increase the lag in your swing and bring the clubhead down on a slightly more narrow track, like the pros do, try the following exercise. Assume your address position with your golf bag standing up several feet behind you, on your stance line. Swing the driver back until the shaft is parallel to the ground—your left hand should be just outside your right foot and the clubhead mere inches from your bag [figure 6-12]. Both arms should be fairly straight, so that the clubhead extends a long way from your body. Swing the clubhead up to the top slowly and then down, stopping when the shaft is parallel to the ground and your hands are in front of your right leg. At this point, the clubhead should be well inside the bag, on a much narrower track than it was on the backswing. This is the wide-to-narrow piece that all long hitters have in common [figure 6-13].

figure 6-13

WIDE-TO-NARROW DRILL II

Assume your normal setup with a driver but no ball. Sole the clubhead about a foot or two forward of its regular starting position [figure 6-14], and then drag the sole back along the ground for several feet, until the club comes off the ground [figure 6-15]. As the club swings up, try to extend the butt end of the club as far away from your body as possible, without moving your head back away from the target [figure 6-16, opposite page]. Before the clubhead comes to a complete stop, start shifting more weight to

figure 6-14

figure 6-15

your left foot and swing your hands and arms down in the opposite direction, toward the imaginary ball. Feel as though the butt end of the club is making a beeline toward the ball. The pulling force created by your weight moving in the opposite direction that the clubhead is traveling should increase the lag and the cocking angle between your left forearm and the clubshaft [figure 6-17]. Repeat the drill several times, then put a ball down and see just how much more distance you create with this additional lag.

figure 6-16

figure 6-17

figure 7-1

Dustin Johnson has a hip-to-hip shoulder differential of 71.3° at impact, one of the main reasons why he hits the ball so far.

IMPACT

CHAPTER 7

The Other X-Factor

UP UNTIL THIS POINT, everything you've read in this book has been geared toward achieving one thing—a good impact position. You've been taught to swing the clubhead back with a flat left wrist because that gives you the best chance of returning the clubface square at impact; you maintain your original spine inclination because it allows you to swing your hands and club down into the ball on the proper inside path; and you shift your center of gravity forward as fast as possible on the downswing because that's how to best maintain lag and increase your clubhead speed. Every move you make, including how high you tee the ball, has its fingerprints on what the clubhead and your body are doing at impact.

Keep in mind that with the driver, the ball is sitting on a tee. This has a huge influence on where your body and club should be at impact. In a good impact position, your hands should be slightly ahead of the ball and your weight left, on your front foot. Your spine angle should remain relatively intact (there are some slight variations to each, which I'll get into shortly) and your body should be circling to the left, with your hips pointing left of the target and your shoulders about to square up to the target. As for the club, the head should be approaching the ball from the inside (0°–3°) on a slightly ascending path, with the face square. The handle should remain down with the shaft leaning slightly forward, almost in a straight line with your left arm.

Virtually all Tour players arrive at the impact position described above, no matter how different their swings may appear. Why? Because they know how to get their hands and club on-plane on the downswing, and their sequencing is so much better than the average golfer's. They also use the rotation of their body much more than the average slicer does, which is why they don't have to rely so much on timing to hit the ball far. In a nutshell, they make fewer mistakes than the average golfer does, and thus they don't have to make as many in-swing compensations.

In this chapter, I'll take a closer look at some of the biggest mistakes, or power leaks, that can affect your impact position. I'll also explain once again the role that the hips and shoulders play in creating power, and how the pros are able to make the ball explode off the face like it's being shot out of a cannon.

▌ TORQUE POWER

In the late 1990s, the term "X-Factor" was introduced into the golf instruction vocabulary, and it has been a mainstay ever since. By definition, the X-Factor refers to the separation in shoulder and hip turns at the top of the backswing—the more you turn your shoulders relative to your hips while maintaining your spine angle, the faster your body unwinds and the more clubhead speed and power you're capable of generating (see Chapter 5). Think of a spring: The more you wind it, the faster it unloads. Ideally, you want a shoulder-to-hip-turn ratio of 2:1, with the shoulders rotating at least 90° and the hips no more than 45°.

What few people realize is that there's another X-Factor, this one occurring at impact, which has an even bigger influence on how far you hit the ball. I like to refer to it as the "Torque Factor." The same definition applies—the wider the differential between the shoulders and the hips, the faster your body turns—except that the roles are reversed. With the Torque Factor, you want your hips out-turning your shoulders by a

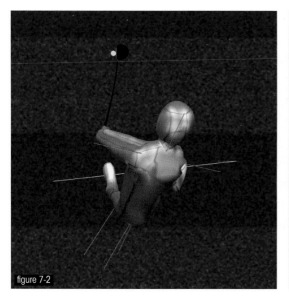

figure 7-2

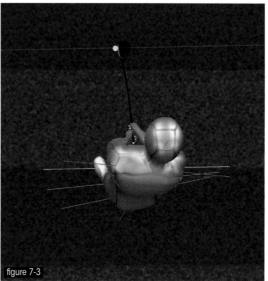

figure 7-3

sizable margin. The wider you make that gap, the more stored-up energy and clubhead speed you're able to release through the ball.

One of the biggest Torque Factors in professional golf belongs to Dustin Johnson, who has a hip-to-shoulder differential of 71.3° at impact [figure 7-2]. Johnson's hips are 59° open to the target line, while his shoulders are still closed (-12.3°). This is ideal: You want your hips to be open as much as possible while your shoulders are still within a few degrees of your original stance line, or toe line [figure 7-3]. By contrast, the average over-the-top slicer's hips are open only 11.4° and his shoulders closed -5.6°, a differential of 17°—or about a quarter of the size of Johnson's Torque Factor [figure 7-4]. That explains why the average slicer drives the ball about 100 yards less. The gap between the hips and shoulders is too small, and thus there's virtually no built-up energy available to release through impact.

Johnson, because of the differential between his hips and shoulders and the amount of torque this produces, unwinds through the ball with a tremendous amount of rotational speed. His hips are turning sharply to the left while his shoulders are still playing catch-up, causing the clubhead to be pulled through like a slingshot to create an extra burst of

Left: Dustin Johnson's X-Factor (the differential between the shoulders and hips) at impact is 71.3°, one of the largest on the PGA Tour. The wider this gap, the faster the body rotates and the more clubhead speed and power you produce [figure 7-2]. Right: The average over-the-top slicer has a hip-to-shoulder differential of only 17°, substantially smaller than Johnson's and most Tour pros [figure 7-3].

energy through impact. Johnson's average clubhead speed (112.6 mph) and initial ball speed (170 mph) are off the charts, which is why he hits the ball as far as he does. In fact, Johnson had one drive that measured 463 yards in 2011, the longest on Tour that year by 44 yards. In stark contrast, the clubhead speed of our average over-the-top slicer (with driver) is 95.7 mph, or 17 mph less than Johnson. Since every mile per hour of clubhead speed equates to roughly 2.5 yards, that's roughly 43 yards he's giving up to Johnson—the difference between hitting an 8-iron and a 4- or 5-iron into the green. You can add another 40 or 50 yards to that number, too, since our average slicer's ball speed is also 35 mph less than Johnson's, due to differences in club speed, how centered contact is, and other launch factors.

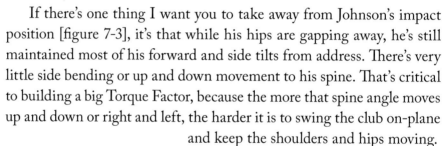

If there's one thing I want you to take away from Johnson's impact position [figure 7-3], it's that while his hips are gapping away, he's still maintained most of his forward and side tilts from address. There's very little side bending or up and down movement to his spine. That's critical to building a big Torque Factor, because the more that spine angle moves up and down or right and left, the harder it is to swing the club on-plane and keep the shoulders and hips moving.

That's what happens to the average over-the-top slicer. Because the shaft is coming in too vertically (i.e., out to in), he has to stand up to avoid hitting the ground a foot behind the ball. As a result, it becomes very hard to turn, which is why he has such a limited Torque Factor. It's as though the shoulders get stuck because the body is unable to continue to turn. Most slicers come up 10° to 15° in spine angle, whereas the average Tour player raises up about half as much. Because his spine tilt is virtually the same and his path (i.e., in to out) is good, the Tour player is able to rotate his body in a circle

Johnson's forward spine tilt (30.5°) remains relatively close to what it was at address, despite the enormous rotational speeds his body generates. This allows him to swing the clubhead down to the ball from the inside, on the ideal power path.

figure 7-4

and whip the clubhead around, whereas the average slicer isn't able to rotate at all and has a hard time compressing the ball with any force.

To increase your Torque Factor at impact and boost your clubhead speed significantly, you must first learn to swing the shaft on-plane during the downswing, on a slightly inside (0°–3°), circular path. As soon as the club drifts off-plane and the shaft gets too vertical, you start losing clubhead speed. The handle of the club rises and so does your body, and the hips and shoulders stop turning.

About 90 percent of all recreational golfers throw the club off-plane in the transitional phase of the downswing. Their first move is to lunge forward with their right shoulder, which keeps the weight on the right side too long and kicks the hands out toward the target line. I call this the golf swing's "death move." The minute it happens the club is going to come over the top and you'll have to make some compensating move to get it back on-plane. Only the game's best players are able to do that. To learn how to better swing the club on-plane, see the "Trace the Shaft" drill at the end of this chapter on page 88.

The second step to increasing your Torque Factor is to improve the amount of clubhead lag on your downswing. The longer your wrists remain hinged on the downswing, the longer the clubhead lags behind your hands, which sets up a powerful slinging action through impact as the wrists and shaft begin to straighten out. As you can see in the screen capture on this page, the clubhead on the over-the-top slicer's avatar has already passed the handle and is slowing down [figure 7-5]. This early release, or "casting" of the club, is a major power drain and often contributes to a weak, left-to-right slice. In a good impact position, the handle remains low and slightly ahead of the clubhead, which is about to reach its maximum speed.

This is an impact position you want to avoid—handle high, weight back, clubhead forward of the hands [figure 7-5]. If the clubhead passes the hands prematurely, it means it's actually slowing down as it contacts the ball.

figure 7-5

The final step, provided that you can consistently deliver the club-head on-plane, is to feel the stretch between your hips and shoulders as you unwind through impact. This will not only increase your lag, it will also help you deliver the clubhead to the ball from the inside. One thought that many Tour players have is to keep their back to the target for as long as they can. This delays the shoulders just long enough to prevent them from moving out over the top on the downswing, and forces them to play catch-up. That's the feeling you ultimately want to have—that of your shoulders and right side firing through impact after your hips and knees have already cleared. Now that's power!

DRILL: BALL UNDER BRIDGE

Take two range baskets, turn them upside down, and place them several feet in front of you so that they're just out of reach of your swing. Spread them out and lay an old shaft across them horizontally so that it forms a bridge [figure 7-6]. With your 7-iron, hit some half-swing shots with the vast majority of your weight firmly planted on your left heel. The object is to hit each ball under the bridge, which can only happen if your weight is left, the shaft is leaning forward (toward the target) at impact, and your hands are on-plane [figure 7-7].

figure 7-6

As you're working on this drill, note how close your arms are to your chest as you swing down, and how there's no separation of the elbows. The handle stays down—it doesn't go vertical—and your left foot doesn't move. The latter is key: The minute you pick that left heel up off the ground, the shoulders start to open and the shaft moves out toward the target line. You find this a lot with over-the-top swingers, because their weight is still back on their right foot as they start the transition. What you see is the left foot coming up off the ground and actually moving toward the right foot, which puts the golfer in a very weak hitting position. There's no way this can happen if most of your weight—about 90 percent—is already on your left foot as you start down. With the help of this drill, you should immediately sense when your weight is forward enough and you're in the best position to start your downswing.

figure 7-7

Throughout this book you've heard me repeatedly mention the importance of maintaining spine angle, yet if you look at Johnson from a straight-on angle at impact [figure 7-8], you'll see that his spine has actually backed up 8° from where it was at the top of his swing. His forward tilt toward the ground has also come up some, from 34.1° to 30.5°. Both changes are necessary in order to hit the ball on the upswing and avoid contacting the ground.

As Johnson swings down, his left hip pushes forward and up [figure 7-8], and his butt tucks under his hips as they clear sharply to the left (refer back to figure 7-4). This is what causes the spine angle to back up and raise slightly. If it didn't, the clubhead would more than likely bottom out behind the ball or come in too steeply, and he wouldn't be able to hit up at the ball. The average golfer also increases his side tilt (from 0.3° to 6.9°), but struggles to get his weight forward, onto his left foot,

Left: Dustin Johnson's spine has backed up considerably (8°) from where it was at the top of the swing, which allows him to hit up on the ball and avoid contacting the ground [figure 7-8]. Right: Our average over-the-top slicer is practically standing up straight (17.4° forward tilt) at impact. This is due to an improper weight transfer and an overly steep downswing path [figure 7-9].

figure 7-8

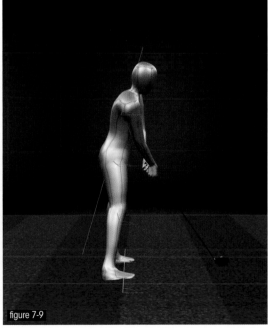

figure 7-9

DRILL: POST UP THE LEFT LEG

Torque on the downswing is created by unwinding your lower body against your upper body, keeping the shaft on-plane and your spine angle constant from the transition through impact. One way to experience this "Torque Factor" stretch is to practice the following drill. Take your normal stance with a 6-iron and stick a shaft in the ground just outside your left hip. Make several practice swings without a ball, getting a feel for how your left leg straightens and posts up against the shaft at impact [figure 7-10]. Now make several three-quarter swings with a ball, keeping your shoulders square to the target line and your right heel banked at 45° (up on your toes). Provided you're able to keep your hips forward and your spine angle intact, you should be able to turn your hips through more efficiently, widening your Torque Factor and increasing your clubhead speed [figure 7-11].

To create maximum power, you can't have your hips spinning open prematurely and your weight staying back at impact. Your hips need to move forward (think of them bumping an imaginary shaft) and your belt buckle must stay high. That snaps the left leg straight—a move common with all long hitters—which in turn opens the hips. If your hips open without being forward it will result in an over-the-top downswing and no torque.

figure 7-10

figure 7-11

because his initial downswing path is too steep (i.e., out to in). Consequently, he has to stand up even more and his hips and body stop rotating [figure 7-9].

There's another reason why Johnson and other Tour players push forward and up with their hips, and that's to create additional power. At the start of the downswing, Johnson's forward spine tilt actually drops 0.3° toward the ground before rising slightly upward as the clubhead approaches impact. What he's doing is using the resistance of the ground—like a broad jumper about to take flight or a sprinter pushing off the blocks at the start of a race—to spring his hips forward and generate more upward momentum at impact. Raising the belt line also straightens and firms up the left side, creating a solid post for the hips to rotate around. This frees up space for the hands and arms to swing down in front of the body, producing even more clubhead speed [figure

figure 7-12

7-12]. Even with his belt going up and his butt tucking under, Johnson still maintains his original spine angle very closely, because this is the only way to keep the club on-plane.

HIP POINTERS

One of the more important body statistics that the MAT-T system is able to track is hip angle. This is the angle created by the movement of the hip joint. Another way to look at it is that when you bend forward, there's a differential between your femur (i.e., thigh bone) and pelvis. That's your hip angle.

At impact, the Tour player's left hip (in silver) is significantly higher and more forward than the average slicer's (in red). The Tour player's left hip is also rotating around to the left, or clearing, while the average golfer's lead hip is barely rotating at all.

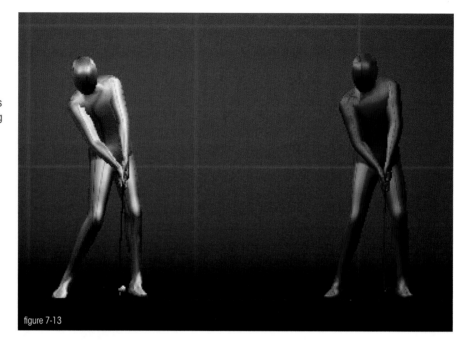

figure 7-13

If you look at the side-by-side comparison (see figure 7-13 of the Tour composite golfer [left, in silver] and the average out-to-in slicer [right, in red] at impact), you'll clearly see that the Tour player's left hip is higher and more forward than the slicer's. His left hip is extending upward and rotating to the left at the same time, while the slicer's left hip remains down and square to the target line. That's why the slicer has a significantly higher left hip angle (31.7°) than the Tour pro (19°): While the slicer is standing up at impact, his hips have not pushed forward and up fast enough on the downswing, so there's more bend to his waist—i.e., more angle. He's all jammed up, whereas the Tour player has plenty of room to extend and accelerate his arms because his left hip has cleared out of the way.

The only way your left hip is going to be able to create this space is if your weight is already on your left side at impact. As your center of gravity starts to shift toward the target, your left hip begins to push forward and up, which has to happen if the clubhead is to avoid the ground and hit the ball on the upswing. To promote this forward, upward movement of the left hip, try the following drill.

DRILL: PICK YOUR BELT UP

Set up with your head resting against a wall and your hips and shoulders square to the wall. Cross your arms over your chest, and rotate your body into an imaginary impact position, rotating your hips to the left while keeping your shoulders square to the wall. Use the ground to help you push your belt buckle forward and up as you clear your hips [figure 7-14]. Try and lift your belt buckle as high as you possibly can. Hold the position for several seconds: You should feel a big-time stretch in your trunk, almost like your hips are being screwed into the ground, and your belt buckle should be higher than it was at the start of the exercise. There should also be no weight whatsoever on your right heel. Repeat several times, posing at impact so you can feel where your left hip and your body should be in this pivotal position [figure 7-15].

figure 7-14

figure 7-15

Take your normal stance with a 6-iron and then stick a shaft in the ground roughly three feet behind the ball, along your target line. Angle the shaft away from the target line at 45°, so that the handle points behind you. (Note: You can substitute an umbrella if you don't have a shaft.) Swing the club down from the top very slowly, trying to match the path of the clubhead to that of the shaft. Imagine that the shaft is a plane board extending above your right shoulder, and pull the handle down so it's tracing along the imaginary board. When the clubhead is just below waist height on the downswing, the club should glide along the shaft [figure 7-16].

This drill trains your hands to swing down and backward at the start of the downswing, not out toward your head and the target line, which puts the club on the correct inside path. It also encourages you to maintain the same spine angle you had at address. When you're in this posture, your body can turn a lot faster than it would if your butt came in and you stood up straight. The minute the body starts to stand up, that's when it stops turning and you lose the ability to create more clubhead speed and power.

figure 7-16

THE "IN" PATH

I touched on it earlier in this chapter, but I cannot stress enough the importance of swinging the clubhead from the inside on the downswing. An out-to-in swing path is like a blackout when it comes to generating power, because when the clubhead swings down from the outside your body cannot rotate as fast as it can when the head is traveling with it—

i.e., on a circular path from the inside. Once the shaft gets too vertical, your body has to extend early and your trunk rotation virtually comes to a halt. That's what we're trying to avoid, an early extension of the body. If you can avoid coming over the top and can maintain your spine angle through impact, your body can rotate in a circle and the clubhead can swing down from the inside, which is the fastest and most efficient way to swing the golf club.

The average Tour composite golfer swings 1.2° from the inside, whereas the average slicer is a -5.1° from out to in. The slicer is not only cutting across the ball, he's swinging down (-0.2°), which leads to a higher spin rate, more slice spin, and a lot fewer yards. Work on the drill "Trace the Shaft" (opposite page) in practice to help you shallow out your downswing plane and keep the clubhead tracking from the inside.

figure 8-1

Sergio Garcia's arms are fully extended and the clubhead is far away from his body in the post-impact position.

FOLLOW-THROUGH

CHAPTER 8

Extend for Power

MAKE NO MISTAKE: THE most important position in the swing is impact. What happens at this decisive "moment of truth" dictates the direction, trajectory, and distance that the ball travels. When the ball leaves the clubface, there's nothing you can do to change the result. It's gone. But that doesn't mean you should ignore what's happening with your swing after impact, because while your follow-through might not save an errant tee shot, it can tell you a lot about what's going on at impact or just prior to it. Most importantly, it can inform you as to what you're doing wrong.

Many of the flaws you see after impact (i.e., a chicken wing or a reverse "C" finish) are a direct result of what occurs before or at impact. For example: If you find that most of your weight is still on your back foot in the finish position, you can rest assured that you didn't get your weight to your front side fast enough on the downswing. Chances are you came down over the top of the ball as well. No Tour professional plays the game with their weight on their right leg on the downswing and at impact.

On the plus side, you can improve your pre-impact and impact positions by working on your follow-through. Impact happens so quickly that you need to slow things down and feel some simple moves to put you in a better position to strike the ball. In this chapter, I'll take you through a few of these moves and provide you with some drills to not only improve your impact position, but make you finish like a pro as well.

■ HOT OR MILD? THE DREADED CHICKEN WING

At impact, you'll see a lot of golfers, generally slicers, shortening their arms to avoid hitting the ground early. The left elbow bends and then collapses even more shortly after impact, creating an arm that resembles a chicken wing [figure 8-2]. Most Tour players will have a little bit of bend to the left elbow in the mid-through position (18.9°)—when the shaft is parallel to the ground—and virtually nothing at all at impact

figure 8-2

(11.3°), whereas your typical over-the-top slicer will have more than three times as much in the mid-through position (61.5°) and twice as much at impact (20.5°).

There are a number of ways to chicken wing, from helping the ball up in the air (the spine angle comes up too much at impact) to hanging onto the club too tightly, but the primary culprit is a poor transition. If you do not get your weight to your left side fast enough on the downswing, the clubhead will take a steep, out-to-in approach into the ball, forcing you to stand up and bend your arms to avoid hitting the ground several inches behind the ball. This early extension not only makes it difficult to compress the ball and control the clubface, but it also causes the body to stop turning in a circular fashion, which sacrifices speed and power. The Tour player's hips don't stop turning until the swing is complete, but the average over-the-top, chicken-wing player has very little hip rotation prior to and after impact. If the hips stop rotating, there's little room for the arms to extend and they thus have to separate even more, which increases the size of the chicken wing.

■ HOW TO SNAP THE CHICKEN WING

If you look at a side-by-side comparison of Sergio Garcia [figure 8-3] and our average over-the-top slicer [figure 8-4] in the mid-through position, you'll notice that the grip end of Sergio's club is much farther away from his body than that of the average golfer's, as is the clubhead. The latter cannot stretch any farther away from his body, an indication that his weight is forward, his hips are still clearing (creating room for his arms to extend and swing forward), and the club is on-plane. All signs here point to a very big, successful drive.

If you come over the top, your hands will get pulled in close to your body on the follow-through side. As a result, you won't have the room to extend your arms, thus narrowing the swing's arc (i.e., the distance between the clubhead and your body) and significantly reducing your clubhead speed. Much like the backswing, you want to create as much width as possible between the butt of the club and your body, because this gives

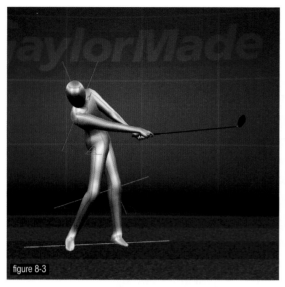

figure 8-3

figure 8-4

The average slicer's left arm resembles a chicken wing in the mid-through position, while Sergio Garcia's left arm is almost fully extended and the clubhead is stretching away from his body [figure 8-3]. When you swing across your body, from out to in, you pull the handle into your body, causing the left arm to fold [figure 8-4].

the clubhead more distance to travel and accumulate speed. You simply can't generate much speed or power if you're swinging across and into your body.

So just how do you get this full extension—and, in the process, avoid the chicken wing? First, you have to get your weight to your front side as quickly as possible during the transition from backswing to downswing (see Chapters 6 and 7 for some exercises to help you with this weight transfer). This is the only way to maintain your original spine inclination and keep the shaft on-plane. It will also give you a much better chance at controlling the clubface. If you start to stand up on the downswing, the handle of the club will also rise, opening or closing the face. Most of the time the face will be closed to the target but open to the path of the clubhead, causing the ball to start left of the target line and then slice to the right.

The most common mistake that golfers make is that they shift too much of their weight onto their right side on the backswing, instead of staying centered over the ball with the right amount of shoulder tilt. This excess lateral movement, known as a sway, gives people the sensation of being able to shift their weight back to their front side better (on the downswing). Unfortunately, it often has the opposite effect and causes a lack of forward movement, because your spine has to back up in order for you to get behind the ball and hit it on the upswing. Make sure your spine remains in its original inclination as you turn back over your right leg; this will put you in the best position to move forward on the downswing.

Provided your address posture remains intact, you should be able to keep rotating your hips, which is what gives your arms the space to swing down in front of your body and extend into the follow-through. If you lose your posture and stand up at the start of the downswing, it's very hard to turn and swing the club on the proper plane, which is on a circle. That is the fastest, most powerful way to swing a golf club. If you were to trace the arc that Sergio's clubhead takes during the swing, it would resemble an oblong, almost egg-shaped oval, with the clubhead at the widest, outermost portion of the egg in the mid-through position [figure 8-5]. This is because his hands and club are moving at right angles to his

figure 8-5

spine in a circular manner around his body; they're not moving in an up-and-down, linear fashion.

Lastly, if you want to break the wing, you need to keep your head behind the ball at impact, in relatively the same position it was in at address. Sergio's head moves back and down slightly with the driver because the ball is teed up and he's contacting it on the upswing [figure 8-6]. Your head should never move forward of your original address position at impact, as this will cause you to lose your spine angle and bottom the clubhead out too soon, necessitating the bent-arm, chicken-wing position.

The chicken-wing syndrome afflicts many golfers and just about every over-the-top slicer. Smooth out your transition and work on the following three drills, and you should improve your impact position and overall ballstriking.

figure 8-6

DRILL: PUNCH AND HOLD

Lodge a rolled-up towel or small ball (Nerf ball, volleyball) between your forearms, and make some waist high–to–waist high swings with a 7-iron, braking your arms immediately after impact [figure 8-7]. Feel as though you're applying pressure to the towel with your forearms as you swing back and through: This will push the handle forward and keep the hips moving forward, so that they can continue to rotate to the left and pull the club with them (below).

Most over-the-top slicers have a hard time stopping their swing in this drill, because they throw the clubhead early from the top. As a result, the arms have to shorten and separate so that the head doesn't hit the ground, which is what leads to the chicken wing. Hitting punch shots with the towel encourages you to swing the club down on the proper inside path, because that's what keeps the arms together and prevents the towel from dropping out. It's much harder to do that if the club is swinging down from out to in.

figure 8-7

DRILL: PULL THE ROPE TIGHT

Tie a rope around your left shoulder and then fasten the other end to the shaft, just below the grip. Starting from the halfway-down position on the downswing, make some 50 percent swings, stopping when your hands are about waist high on the follow-through. If your arms extend through impact, your elbows will stay close together and the rope will remain taut [figure 8-8]. However, if your arms start to fold, your left elbow will collapse into a chicken wing and the rope will have some slack to it [figure 8-9]. Make several practice swings, keeping the rope tight.

    Note the triangle formed by your arms and shoulders and how little separation there is between your elbows. Both arms are extending out toward the target. This is the feeling you want to have after impact, almost as though you're trying to reach out and touch the target. If your arms are pulling into your body, then you can't create any clubhead speed; they have to be swinging out toward the target.

figure 8-8

figure 8-9

figure 8-10

THE FINISHED PRODUCT

At the completion of the swing, you should feel as though you could hold your pose for several seconds. From straight on, your body should look like a tall statue, the left side virtually hidden by the right [figure 8-11]. The belt should point at the target or slightly left of the target, and your right heel should be banked. The toes on your right foot should touch the ground but only for balance; there is no weight on this foot. All of your weight should be on your left side.

This is unquestionably the biggest difference between the Tour player and the over-the-top slicer in this position. The weakest hitters almost always have their weight on their back foot at the finish. Sometimes they'll fake their follow-through pose in an effort to look like a player with a nice, balanced finish, but their center of gravity is usually well behind the ball, a result of not getting their weight forward fast enough during the transition. When you couple this back foot position with an aggressive lateral body slide (toward the target), you get the classic reverse "C" finish, in which the spine arches awkwardly backward [figure 8-12].

The reverse "C" is not as common as it used to be, but nevertheless, finishing with your weight on your back foot will not only diminish your power, it also can cause you some lower-back issues over time. The key to straightening out the "C" and finishing in a more upright, balanced position is, once again, to get your weight to your front foot faster on the downswing. The earlier you do this, the more hip rotation you'll create and the more clubhead speed you'll produce. Try to get your belt buckle to point slightly left of the target at the completion of your swing, not toward right field (i.e., right of the target). This will encourage more hip rotation. Your head should finish forward enough so that it's almost over your left foot. The power hitter's head is back at impact, but then it releases up to finish forward of its original address position. If your weight is back, this can't happen.

The longest hitters finish tall and in balance [figure 8-11], whereas the weak hitters struggle to get their weight to their front foot on the downswing, which often results in a classic reverse "C" finish [figure 8-12].

figure 8-11

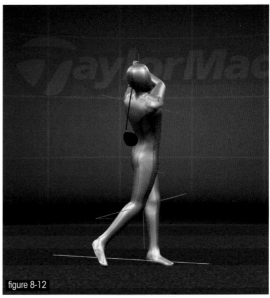

figure 8-12

Be careful not to lift and turn the left foot on the downswing. A lot of golfers feel as though they need to pick up the left heel to facilitate a better turn—and more power—but all this does is prevent them from getting to their left side fast enough. As a drill, hit a few drives on the range and hold your finish for a count of five seconds. (I'll often say the phrase, "Hold it, hold it," with my students.) Make sure your weight is forward and over your front foot, your belt buckle points as far left as possible (don't fake it by turning your hips after you've come to a stop), and your right foot is banked [figure 8-13]. Someone standing behind you should see the cleats on the bottom of your right shoe. Pay close attention to your left foot as well, as you want to keep this as stable as possible; you don't want to turn this foot once you've started your downswing.

If you reach this forward, balanced position, chances are you put a good swing on the ball. But that's not always the case. Many golfers try to look the part and put some body English on their finish position to make up for a bad weight shift. If you come late to the party—i.e., you don't move your center of gravity forward fast enough—it's not going to matter what you do after impact to shore up your finish position. Most golfers have never experienced what a proper finish position feels like.

figure 8-13

That's why the above drill is such a good one to practice. If you're able to repeat this finish position, and you know what it feels like because you've done it so often, then your pre-impact and impact conditions are bound to be good, and you'll hit the ball with much more authority.

Dustin Johnson's tremendous flexibility and athleticism allow him to routinely hit drives more than 200 yards.

OVERCOMING YOUR BODY'S LIMITATIONS

THROUGHOUT THIS BOOK, YOU'VE heard me talk about the importance of establishing good posture at address and maintaining your spine angle throughout the swing. This is something that all of the long hitters on the PGA Tour do: They stay within a few degrees of their original spine inclination through every phase of the swing. Their hands and arms rotate in a circle around this fixed axis point until well after the ball is released, which is what allows them to swing the club on-plane and generate maximum clubhead speed and ball speed.

If you want to maintain this constant spine angle through eighteen holes of golf, you can't expect to just show up on the first tee, wind your body up, and go. The golf swing places a ton of physical demands on the body, particularly on the lower back and core regions that are so critical to maintaining this spine angle. When hitting a driver, for example, the compressive, shearing forces acting on your lumbar spine are eight times that of your own body weight. That's 1,600 pounds for someone who weighs 200 pounds. Your body must be prepared to handle such forces, or it's likely to break down—resulting in injury—or to compensate in such a manner that limits your ability to rotate around your spine and swing efficiently.

In years past, most golfers failed to consider the possibility that their body, and not their swing, could be the primary cause behind their poor performance on the course. If they were struggling mightily off the tee, they'd pin it on something mechanical, and then look for a quick fix by reading the latest golf instruction magazine or scheduling a lesson with their local teaching professional. Or they'd blame it on their equipment and go out and buy a new driver. But now, with Tour players looking more like Olympic athletes and golf-specific fitness trainers sprouting up at facilities all over the country, there's a greater appreciation and awareness for the role that the body plays in the swing. Many golf instructors can now give you a quick functional assessment of your body, which allows them to pinpoint any physical deficiencies that might be adversely affecting your swing.

Most of the big power leaks I've identified in this book—a flat shoulder turn, an early extension of the spine on the downswing, an over-the-top downswing path, a chicken-wing finish, etc.—are all associated with postural changes. All of these faults have common physical denominators that can cause you to lose your spine angle and forfeit valuable distance and power, says Dr. Troy C. Van Biezen, consultant with TaylorMade Adidas Golf and therapist for several players on the PGA Tour. In this chapter, we'll look at a few of these faults and some of the biomechanical and swing limitations that can lead to them. The following exercises—courtesy of Dr. Van Biezen—will help you build better strength, endurance, and flexibility in these areas so that you can execute each of the moves in this book efficiently.

■ SWING FAULT: REVERSE SPINE TILT (BACKSWING)

The one fault most commonly associated with the backswing is a reverse spine tilt. When this occurs, the spine tilts toward the target at the top of the swing, causing a seesaw effect on the downswing in which the spine then reverses course and tilts away from the target, in the opposite direction. This is the only way that the golfer can get behind the ball at impact. Unfortunately, with all of their body's momentum falling away from the target at impact, it's difficult to transfer much speed and energy into the ball and hit it solidly.

There are a number of physical factors that can cause one to reverse tilt, including a weak core, poor glute stability, decreased thoracic-spine mobility, and a decrease in right hip internal rotation. The right hip plays a crucial role on the backswing, because as you turn back, the lead hip has to open independently of the rest of your body. There needs to be some resistance between your upper and lower body, otherwise there'll be no separation between your hips and shoulders (i.e., the amount they turn) and very limited torque, or X-Factor, on your backswing. The following three exercises will improve hip stability and increase the range of motion in your hips. This is critical in maintaining a stable torso and pelvis so that you can create separation on your backswing without losing your spine angle.

▮ CLAMSHELLS

Lie on your side with both knees and feet touching and a resistance band wrapped around your legs, just below your knees. Make sure the band is fairly taut. Keeping your feet touching and your bottom leg grounded, raise the top leg up toward the ceiling, away from the bottom knee [figure 9-2]. Start with a minimal-resistance band before progressing to a high-resistance band, performing three sets of ten reps. This exercise helps to stabilize the glutes while also adding strength and mobility to your external hip rotators and hip abductor muscles.

figure 9-2

DOUBLE-LEG GLUTE BRIDGE

figure 9-3

Lie on your back with your hands across your chest and knees bent, feet flat on the floor. Engage your glutes and core muscles and slowly raise your butt off the floor until your spine is straight. (Do not hyperextend your back.) Hold this bridge position [figure 9-3] for two seconds and then return to the starting position and repeat. Perform three sets of ten reps to help strengthen the glutes, the primary stabilizers in your lower body. The more solid your foundation is, the easier it is to turn your upper body while remaining in your spine angle.

ONE-LEG AIRPLANE

figure 9-4

Stand on your left leg with your right leg behind you, and your arms forming a T, like an airplane. Bend forward from your hips as though you're addressing a ball, and then rotate your torso to the left and the right, keeping your left foot, knee, and hip in line with each other. As you rotate down, you should feel as though you're trying to touch your toes [figure 9-4]. Perform ten to fifteen rotations on each leg, maintaining good balance throughout the rotation. This exercise will improve your shoulder/pelvis separation and balance and get you used to rotating around your hips while keeping your feet planted on the ground.

■ SWING FAULT: EARLY EXTENSION (DOWNSWING)

I see golfers lose their spine angle most often on the downswing. Their spine straightens and their lower body is thrust toward the golf ball, forcing the torso to elevate through the hitting area and the shaft to come in too steeply. This early extension can be caused by a number of swing (poor transition) and physical factors (tight hip flexors, hamstrings, and calves; weak glutes; poor asymmetry in the muscles and joints around your core). These physical limitations can prevent you from getting into the proper posture (i.e., spine tilts) at address, making it difficult to maintain a constant spine angle throughout the swing.

In order to rotate efficiently around your spine, you must have good flexibility in your hips and shoulders, particularly the hip rotator muscles. If your pelvis is unable to rotate around your lead hip (the left hip for right-handed golfers), then your spine has no other choice but to move up, causing your pelvis to move forward (toward the ball).

Another cause of early extension is an inability to separate your upper body from your lower body as you rotate your torso through impact. Without any clear separation, usually due to reduced thoracic spine mobility, it's hard to keep the lower body and the spine stable. Your core strength is also critical to maintaining a constant spine angle, because these muscles (the abdominals and glutes) help keep your trunk flexed forward throughout the swing.

The following three corrective exercises will address all of the physical limitations mentioned above, and will help improve your stability and range of motion so that you can maintain your spine angle on the downswing and through impact.

■ KNEE-INS

Lie on your back with your arms at your sides (palms down) and your feet outside shoulder-width. Your toes should be pointing straight ahead. Squeeze your knees together without touching while keeping your feet

figure 9-5

flat on the floor [figure 9-5]. Hold for a count of five to eight seconds and repeat, performing three sets of fifteen repetitions. If your knees do touch, spread your feet farther apart and try again. This exercise helps to improve the internal rotation of the hips, which allows the golfer to rotate to the left on the forward swing without losing their spine angle.

T-SPINE EXTENSION + ROTATION

Get down on all fours and place one hand behind your head with the "up" elbow pointing toward the opposite hand. Open your chest to the wall slowly and reach your elbow toward the ceiling [figure 9-6]. Go only as far as your range of motion will allow with minimal movement in the lower back. Try to get all of your rotation from your thoracic spine, which is located in the chest area. Perform three sets of fifteen reps both ways to help increase your T-spine mobility, which will help you to maintain your spine angle throughout your swing.

figure 9-6

A word of caution: This exercise can be very difficult for some people, especially those who work at a desk job and travel a lot. This sedentary lifestyle can lead to rounded shoulders, tight chest muscles and limited T-spine mobility, a common cause of lower back pain since the lumbar spine has to take on an increased role in rotation, something it's not meant to do. If you can't reproduce this movement in a controlled environment like your home or the gym, it will be very difficult to maintain your posture while swinging your driver at close to 100 mph. Don't give up so easily, however: This is a very effective exercise if you stick with it. Perform the exercise in a slow, controlled manner, and don't try and stretch the rotation beyond your limits.

▌ BODY BRIDGE

Start in a normal push-up position except with your elbows on the floor, directly under your shoulders. Both forearms should be touching the ground. Raise up off the floor, squeezing your glutes to activate your core. Try to maintain a straight line, or bridge, from your head to your heels, keeping your shoulder and hip joints aligned [figure 9-7]. Don't allow your butt to sag or lift upward. Hold the position for a count of ten to fifteen sec-

figure 9-7

onds and repeat five more times for one set. Try to do three sets of six reps. This exercise really targets your core and, in particular, your glute muscles. It allows the golfer to have a solid base, so that the pelvis can rotate efficiently around a stable axis (i.e., the spine angle) on the downswing.

▌ SWING FAULT: LACK OF SEPARATION AT IMPACT

At the start of the downswing, your lower body should begin to separate from your upper half. The hips start to unwind, or rotate to the left, while the upper body remains closed to the target. It's not until after impact, usually in the mid-through position, that the shoulders finally catch up to the hips. Based on 3-D kinetic sequence data, the hips are actually decelerating before impact, while the torso and shoulders are

ENERGY TRANSFER AND THE KINETIC CHAIN

To maximize your clubhead speed and power, the ground must be the first link in the chain of energy transfer from backswing to downswing. Too many golfers start down with their arms and shoulders, moving their hands out toward the target line. This forces the clubhead out over the target line, causing you to swing down on a steep, out-to-in path, which produces less distance.

Newton's Third Law of Motion states that "for every action there is an equal and opposite reaction." In other words, for every force applied by one object onto a second, an equal and opposite force is applied from the second object back onto the first. At the start of the downswing, you want to drive your feet into the ground [figure 9-8], which emits an equal force back into your body, says Dr. Van Biezen. This ground force reaction, or GFR, is then transferred up into your legs, pelvis, torso, and arms, and then finally into the clubhead and ball. This would be considered the optimal "kinetic sequence." Data reveals that during the transition phase amateurs apply about 45 percent of their body weight into the ground, while the pros transmit about 90 percent of their body weight.

This transfer of energy moves through what is commonly known as the body's kinetic chain, says Dr. Van Biezen. Each body part acts as a link to the next part, transferring energy through the body as though it were passing through electrical wire. As long as no link in this kinetic sequence is broken, you should pass a tremendous amount of energy from the clubhead to the ball. One physical limitation or weak link can interrupt this chain of energy transfer, causing you to overuse or compensate with another body part to make up for this lack of speed and power. That's why it's critical to undergo a Functional Screening evaluation to determine where the weak links and limitations are. Your workouts can be geared specifically toward these dysfunctions, so that you can correct them and build a more efficient kinetic chain. Most workouts without a baseline Functional Screen will be counterproductive and may cause injury or lack of progress.

On average, Tour pros like Dustin Johnson transmit about 90 percent of their body weight into the ground at the start of the downswing.

figure 9-8

still accelerating. This essentially creates a slingshot effect, with the club-head picking up speed as the shoulders are pulled through.

Our average Tour composite model has a hip-to-shoulder differential of 37.2° at impact, compared to just 17° for our average golfer model. In most instances, the bigger the gap is, the bigger the drive. Unfortunately, most amateurs have a hard time separating their lower body from their upper body, often due to physical limitations such as poor flexibility and a weak core. The following exercises will help to increase the range of motion in your hips and shoulders, as well as to stabilize your lower core so that you have a solid foundation for this separation (also commonly referred to as the impact X-Factor) to occur.

WALKING LUNGE WITH T-SPINE ROTATION

From a standing position, lunge forward with your right leg, contracting your glutes to help stabilize your pelvis. Place your left hand on your right knee and then pull your chest open (to the right) so that your right hand is pointing directly behind you and your right arm is on the same line as your shoulders [figure 9-9]. Keep your hips pointing straight forward; don't rotate them along with your thoracic spine (mid-back). Repeat by lunging on the opposite leg and rotating your T-spine to the left. Perform two sets of ten reps on each leg, holding the end pose for two seconds. As with the previous T-spine exercise, perform the movement in a slow, controlled manner and do what feels comfortable; don't force the rotation. This exercise targets your quads, glutes, internal and external obliques, and hamstrings. It helps to increase your upper-body mobility by improving your lower-body stability.

figure 9-9

▋ MED BALL LIFT

figure 9-10

Start in an athletic stance, with your knees slightly bent, back straight, and hips back. Hold a medicine ball just below the outside of your right knee (the weight can vary from three to eleven pounds, depending on your fitness level), then reach the ball diagonally across your body as though you were trying to touch the corner of a door frame [figure 9-10]. Rotate your hips and chest as you lift the ball upward. Repeat three to five times and then perform the same number of reps in the opposite direction. This exercise helps to improve rotational mobility and core stability, which is critical for creating the separation between the pelvis and shoulders. The core plays a major role in the deceleration process while also acting as the primary accelerator on the downswing, driving your legs and hips toward the target.

▋ RUSSIAN TWISTS

From a seated position on a Swiss ball, walk down into a bridge position so that your head and shoulders rest on the ball. Your feet should be a little wider than shoulder-width apart, and both knees should be bent. Holding a one- to three-pound medicine ball between your hands, rotate your upper body to the left and then the right, keeping your arms straight and your hips level. Rotate back and forth in a very controlled manner, pausing for a second at the end of each rotation [figure 9-11]. Perform three sets of fifteen repetitions. This exercise increases your spinal mobility while adding stability to your lower core. It helps to close the gap between your hips and shoulders as the clubhead approaches impact, transferring maximum energy through the body and the club into the ball.

Note: In addition to his role with the TMAG, Dr. Van Biezen is co-founder of Motion Sports Consulting and the Motion Sports Institute. He specializes in Active Release Techniques, or ART, which is a form of soft-tissue therapy, and has treated athletes in several professional sports besides golf, including the National Football League (NFL), National Basketball Association (NBA), and the National Hockey League (NHL).

figure 9-11

figure 10-1

BUILDING THE PERFECT DRIVER

UNLIKE MOST LAUNCH-MONITOR SYSTEMS, the MAT-T system tracks what both your body and the club are doing, so that you know within a tenth of a degree what the face angle, swing path, up or down path, and lie angle of the club are at impact. Not only that, the MAT-T system informs you how much the shaft deflects (i.e., how much it's bowing) at impact, which is critical when matching the right shaft and clubhead loft to your swing.

The MAT-T system also accurately measures the velocity of the clubhead itself, whereas most other systems guess what your clubhead speed is based on how fast the ball is traveling and how much backspin and sidespin it has. Reflective markers are placed on top of the clubhead, so that the MAT-T system can track where the head is in three-dimensional space and how fast it's moving. (See Chapter 1 for a rundown of how the system works.) As a result, it's able to measure what's happening to the head and what's causing it to do what it does at impact.

From a clubfitting standpoint, this is extremely valuable information: It tells the fitter how much they need to bend for lie angle, how much loft they need to add or subtract, what type of shaft will work best for that particular swing speed and shaft deflection, and so on. And with the adjustability features now present in most drivers, the fitter is able to make these tweaks right away, so that the golfer can see a marked improvement in performance in as little as a few minutes. You don't have to schedule an appointment with a fitter or wait several weeks for a new clubhead or shaft to arrive—a few twists of a wrench and presto, you're ready to go.

Over the past ten years, I've conducted more than three thousand fittings on the MAT-T system, and I can say with confidence that the patrons I'm fitting are walking out much better golfers than they were before they came in. They're definitely longer off the tee: I've seen golfers pick up as much as 50 yards in one fitting session! More often than not, I'm seeing gains of at least 10 to 15 yards with the driver. Adjustability has a lot to do with that, but without the measured data supplied by the MAT-T system, it would be hard to tune the driver correctly to max out performance.

As I mentioned earlier in Chapter 1, the MAT-T system was originally designed solely for clubfitting purposes. It's only in recent years that it's come to be known as an equally effective swing diagnostic tool. On the following pages, I'll take a look at some of the key head and shaft measurements that fitters examine when trying to optimize the performance of each of your clubs for you. This chapter is a must read if you want to max out your power, because the more you know about impact and what factors have the greatest influence over how far (and where) the ball travels, the better prepared you'll be to make it happen.

▌ WHAT CAUSES THE BALL TO SLICE?

My No. 1 job as a clubfitter is to bring the clubface angle and swing path closer together at impact. The more square the face is relative to the path of the clubhead, the straighter and farther the ball will travel. You'll not only contact the ball closer to the sweet spot, which increases your Smash Factor (more on this later), but you'll also impart less spin to the ball, so it'll curve less and roll more. The greater the differential between face and path, the more spin you generate and the more susceptible the ball is to ballooning and slicing.

Consider: Our average slicer's swing path is -5.0° out to in at impact, with a face heading of +0.8° open. The face is open relative to the path of the club by 5.8°, which is going to cause the ball to curve sharply to the right. Any time the face is open relative to the path, the ball is going to move to the right. Now, if I could get the above golfer to close the face 3° at impact, he'd still slice the ball, but it wouldn't curve as much. The ball would start slightly left of the target line (because the face angle at impact is 80 percent responsible for the ball's starting direction) and then bend to the right, but because the face angle and path are closer together, the ball would be more likely to find the right side of the fairway rather than the trees.

Our Tour composite golfer has a swing path of 1.2° from the inside coupled with a face angle of 1.3° open. So while the face is still open relative to the path, it's only by a mere .1°. Consequently, the ball is going to start just right of the target line and barely move at all, finishing perhaps 3 or 4 yards right of the target. It's hard to hit the ball any straighter because the face angle is so close to the path.

If there's one thing I've learned from conducting so many fittings on the MAT-T system, it's that the face angle at impact dictates the starting direction of the ball, *not* the path. A steep down path (i.e., attack angle) can also affect the amount of curvature and slice spin that the ball has (because it creates more spin)—but the face angle's relation to the path has the greatest influence over whether the ball is going to slice or not. Bottom line: If you square the face to the path, you can expect to hit the ball much farther and straighter.

HELP FOR THE SLICER

If your tendency is to hit a giant, left-to-right banana ball off the tee, you can get some relief with today's adjustable drivers. Several of today's models, including TaylorMade's popular R11 driver, have movable weights that allow you to optimize the center of gravity location and produce a more draw-biased ball flight. By shifting the heaviest of the weights toward the heel, the toe becomes lighter, thus making it easier to close and square the face at impact. This is important for the reasons I mentioned above, because the more you square the face to the path, the straighter the ball will travel. If your swing path is -4° out to in and you can close the face a few degrees at impact, your ball won't slice as much and you'll start to find more fairways. There's no better way to significantly reduce the magnitude of your slice than by changing the face angle at impact so it's more in line with the path the clubhead is swinging on.

▌SHAFT DEFLECTION

It doesn't matter if your clubhead speed is 85 or 105 mph, centrifugal force will cause the shaft to deflect, or bend, to some extent during the swing. If it deflects too much, it can negatively affect your club position at impact by adding loft, opening or closing the clubface, or changing the lie angle. The more deflection there is, the harder it is to get the face to do what you're telling it to do, which can cause a significant loss of distance. Too much deflection isn't always bad, because if you struggle with a steep angle of attack you will need it to help get the ball up in the air. But in most cases with faster swingers, too much deflection indicates a need for a stiffer tip section on the shaft to stabilize the head better.

If you were to draw a straight line down the club's handle through

figure 10-2

the center of the shaft and hosel, any movement the shaft makes in any direction relative to this line is what we call deflection. If the shaft moves off this line and is forward of the line, or handle, it's deflecting forward [figure 10-2]; if it moves behind the line it's deflecting backward. The shaft can also deflect downward and sideways. In fact, centrifugal force causes every shaft to deflect downward at impact, no matter what the clubhead speed. This is very important information to the fitter— especially with the irons—since it tells us how much we need to adjust the club's lie angle and face angle. The more the toe of the club digs, or drops down, the more upright we want to make the lie angle; the more the heel digs the flatter we want to make the lie.

The MAT-T system can tell us within a hundredth of a degree how much the shaft is deflecting in a negative (forward of line) or positive (back of line) direction. Typically, the worst deflection occurs with the

If the shaft deflects forward of the handle (see red avatar), loft is added to the clubface, causing the ball to launch too high with too much spin.

figure 10-3

over-the-top slicer who dumps (i.e., releases) the clubhead too early on the downswing. More often than not, this causes the shaft to deflect forward of the handle, which increases loft [figure 10-3]. If the loft presented to the ball at impact is several degrees more than the actual loft of the club, you're going to launch the ball too high and with too much spin.

When I see someone with too much forward shaft deflection, the first thing I typically do is reduce the loft of the head to bring the ball flight and spin down. My next thought is to move the center-of-gravity location closer to the heel to help close the face and bring the launch down. I might also look at making the tip section of the shaft stiffer. The stiffer we make the tip section of the club, the less the loft will increase at impact. However, it's important to note that some people actually need the shaft to deflect more to help get the ball up in the air. It all depends on the attack angle of the club: If the head is coming in too steep, they need the deflection; otherwise, the clubhead will bottom out and hit the ground too early. In this case, we might give the player a softer shaft to increase the bend in the tip section. If your angle of attack is so steep that you're hardly presenting any loft to the ball at impact, you need the added deflection to bring the launch angle up.

■ WHAT SHAFT FLEX IS RIGHT FOR YOU?

Don't get me wrong, the shaft is a very critical component in any driver fitting, but in my experience I've found that adjusting the loft and face angle of the clubhead will influence ball flight more than adjusting the shaft will. Many people think that the shaft is more important than the head, and while the shaft can improve ball flight—and oftentimes does—changing the head will make a bigger difference since it's the only part of the club contacting the ball. The shaft never touches the ball.

I typically fit the golfer for a shaft last because I find it helps to tighten dispersion (i.e., bring your misses closer together) and control spin. The faster the clubhead speed, the more deflection there is and the more important the shaft becomes. This is why you'll see virtually every Tour player using a stiff (S-Flex) or extra-stiff (X-Flex) shaft in their driver. They want to launch the ball at 12° or 13° with low spin, but the ball will go into the stratosphere if the shaft is flexing too much. Pros don't need any help getting the ball up in the air, and they prefer a low- to mid-trajectory ball flight because that's what produces more roll and distance; it's also more effective at cutting through the wind. Conversely, someone with low (sub-70 mph) or average clubhead speed (70 to 85 mph) will probably be best served using a senior (A-Flex) or regular (R-Flex) shaft because it will deflect more, thus increasing loft and trajectory. These golfers typically need to carry the ball farther in the air to get extra distance.

Your angle of attack also plays a significant role in determining the right shaft flex for you. Generally speaking, we give a slower swinger a softer, lighter shaft because they need to generate more clubhead and ball speed to carry the ball farther. The faster swinger needs something stiffer and heavier because they're trying to control the ball flight (i.e., to bring it down). There's a fine balance, though, because sometimes we find a slower-swing player who gets more distance out of a stiffer, heavier shaft because they're hitting up on the ball too much. The opposite is true, too: We'll find faster swingers who need a softer, lighter shaft because their attack angle is too steep and they're launching the ball too low.

In summary, if it's more control you're seeking, you should get fit for a stiffer, heavier shaft; if you're vertically or distance challenged, then look for something softer and lighter. There are exceptions, however, which is why you need to get fit for a driver rather than simply buying one off the rack. Many people only see a clubfitter when purchasing a set of irons, but consider how many times you hit a driver during the course of play. Ten to fourteen times? That makes it a pretty important club. Every swing is different, and in order to get the optimum performance out of both the head and shaft of your driver, you need to get a custom-fit.

■ HIGH LAUNCH, LOW SPIN

Imagine, for a moment, that you're watering your plants at home. The hose only stretches so far, so you need to shoot the water a good distance to reach the plants. If you point the hose too high or too low, the water won't project far enough—too high and it will fall straight down from the sky; too low and it will never get more than a few feet above the ground before petering out. There's an optimal angle somewhere between the two hose positions that will allow the water to carry far enough with enough forward momentum to reach the plants.

This is the analogy I like to use to describe the optimum launch conditions for a driver. The longest hitters on Tour launch the ball fairly high (about 12° of launch) with low spin (mid-2,000 rpm), which creates a piercing, rainbow-like trajectory. The ball carries a good distance in the air and stays at its apex for a long time before slowly descending and hitting the ground running. I typically see golfers with low launch (9° to 10°) and high spin (3,000+ rpm). Like a roller coaster, the ball starts out low but then rises very quickly to the top due to the spin. The ball then falls almost straight down with no roll. If you launch the ball high with too much spin, you'll also get no roll. This makes it very difficult to play in the wind or to get a good feel for your distances.

So how do you get high launch and low spin? First, you have to hit up on the ball (the average up path for our Tour composite golfer is +3°) to get the ball launching high enough; and, secondly, you need to keep the loft down, which you do by leading with the handle. If you look at the animated image of Dustin Johnson from Chapter 7 at impact, you'll see that his hands are well ahead of the clubhead at impact. This means that he's presenting less loft to the ball at impact than the actual loft of his driver (10.5°). Because the loft is down, he's also contacting the ball high on the clubface, which is where the least amount of backspin is generated. A ball hit high on the face will spin much less than one hit low on the face. I liken the optimal high-launch, low-spin drive to a tennis player hitting a forehand topspin shot: the player hits up on the ball but with the racket pointing to the ground, creating a topspin effect. That's essentially what Johnson is doing at impact to make the ball rocket off the clubface.

The best way to counteract too much spin for a slicer is to move the center of gravity of the head closer to the heel, which helps square the face and bring it more in line with the path of the clubhead. The ball will fly straighter, but if the path is out to in it will also go low and left, necessitating an adjustment in loft. From a shaft standpoint, the fitter can also bring the spin rate down by making the tip section stiffer, so that it doesn't flex forward as much.

■ SMASH FACTOR

The MAT-T system also employs a launch monitor to track the ball in addition to the club and the head. Many of these launch monitors keep a statistic called "Smash Factor," which is a measurement of how efficiently you hit the ball—i.e., how square the face is to the path and how close you are to making contact on the sweet spot. The maximum Smash Factor number you can get is 1.50. This is determined by taking a player's ball speed and dividing it by their clubhead speed; therefore, if your ball speed is 135 mph and your clubhead speed is 95 mph, your Smash Factor is 1.42.

The closer your Smash Factor is to 1.5, the closer the face angle is to being on the same line as the path, and the more centered (on the sweet spot) your contact is. Dustin Johnson has an average ball speed of

170 mph and a clubhead speed of 122.6 mph, for a Smash Factor of 1.39. So while he averages well over 300 yards per drive, he still has the ability to hit the ball farther—probably an additional 15 to 20 yards. The best way to bring your Smash Factor and distance up without changing your swing is to have your fitter adjust both the loft and lie angle on your clubhead, which should move your point of impact closer to the sweet spot. You may also want to adjust the length of the club—the average Tour player uses a shaft that's 44.5 inches long, while most recreational golfers play one that's a full inch longer, at 45.5". Sometimes you can get more distance by going to a shorter shaft, because then it's easier to control the clubhead and find the sweet spot, which will increase your ball speed and Smash Factor.

■ THE AGE OF ADJUSTABILITY

Most of the big equipment manufacturers today, including TaylorMade, offer adjustability features in one or more of their drivers. This gives you the ability to serve as your own clubfitter by adjusting for lie angle, loft, and face angle to optimize your ball flight for your desired trajectory and shot shape. For higher-handicappers, the adjustability allows for some shot correction so that you can keep the ball in play more consistently and maybe even pick up a few yards.

In my experience, the adjustability dimension has significantly upgraded the performance of drivers over the past few years, to where I can consistently get someone 10 to 15 more yards without changing their swing. With the slicer, I can move the center of gravity location closer to the heel to help them square up the face, and vice versa for the hooker of the ball. I can also adjust the head's loft without bending the face angle or lie angle of the club, and can alter the face angle simply by tweaking the adjustable hosel.

TaylorMade's R11, which was the No. 1-played driver on the PGA Tour in 2011, has three dimensions of adjustability that allow you to adjust loft, face angle, and flight path independently of each other. Movable Weight Technology (MWT), which was available as early as 2007 in the R7 driver, allows you to position the driver's center of gravity closer to the heel or toe by moving the heavier of two screw weights to the preferred location. Moving the weight toward the heel creates more of a draw-bias, which is ideal for those players who struggle with a left-to-right ball, while moving the weight toward the toe promotes a fade-bias.

The adjustable sleeve on the hosel of the R11 might be the most interesting of the three features. The sleeve has a 1° bend to it, so that when you move it around it changes the loft of the head in ½-degree increments. And at the same time it's changing the loft, it also adjusts for lie angle. This Flight Control Technology (FCT) gives you the ability to increase or decrease the loft by 1°, thus influencing both the launch angle and spin rate of your shots. Finally, there's the Adjustable Sole Plate (ASP) on the bottom of the club, which is how you adjust the face angle and look of the club at address. You can set the face to an open (+2°), neutral, or closed (-2°) position, and do so without changing the head's loft.

Many golfers don't take advantage of this adjustable technology because they're wary of change or find the features to be too confusing. But they're really missing out on a tremendous opportunity to increase both their power output and accuracy off the tee. If you haven't demoed one of these adjustable drivers yet, I suggest you do. A fitter can adjust the settings to your liking, so that when you do find a loft, lie angle, and face combination that works best for your swing, you don't have to change anything again. And should you buy the driver and find that you do want to tweak it later, most pros at your home club will have no problem adjusting it for you, free of charge.

■ WHY GET FIT FOR YOUR DRIVER?

If you've read this far, you undoubtedly understand the value of getting fit for a driver. You wouldn't buy a car without first researching it and taking it for a test drive, and you should approach any club purchase the same way. There's a saying, "It's not the arrow, it's the Indian," but name me one Indian who can hit a target from 50 yards away with a crooked arrow. You need a straight arrow to hit the target, just as you need a driver with the right clubhead-and-shaft combination to hit your targets more consistently.

To find the TaylorMade Performance Lab and MAT-T system nearest you, go to taylormadegolf.com.

ACKNOWLEDGMENTS

A person who was very important to me once said, "In some quiet way, the expression and feelings of gratitude have a wonderful cleansing or healing nature. Gratitude brings warmth to the giver and the receiver alike." In this spirit, I would like to acknowledge the many people who have helped with this book and influenced me in my crazy, bizarre career.

I would like to start by thanking Wes Neff, who doubles as my brother and literary agent. (The former is more difficult.) He really pushed me to put this project in motion, and without him, it would have never left the launching pad. In addition, I would like to thank Gotham Books for publishing this book and giving me the opportunity to share my passion with all of you.

The person who has felt the most pain in this process is my writer, Dave Allen. For those of you who know me, you know how hard it is to get an ADD nut-job like me to focus on anything for more than one minute, so you can imagine how much patience Dave has had to have with me. He is a great writer and very detail oriented, which is exactly the type of person I needed to have working with me.

TaylorMade Golf has been an awesome partner to me. They have supported me for a long time now with the MAT-T system and the development of the TaylorMade Performance Labs. Will Miele has made a huge impact on the growth of the TaylorMade Performance Labs and MAT-T system. Ian Wright created the Tour composite avatars for this book, what I like to refer to as the "Holy Grail" of swing sequences. He's a really smart guy. I've truly enjoyed working closely with him and all of the clubfitters, employees, and teachers at TaylorMade and the Taylor-Made Performance Labs. I've learned a great deal from all of you.

I would also like to thank Dr. Troy Van Biezen and Mike Malaska for their contributions to the book. These two men are at the very top of their respective fields, and I'm so grateful to them for lending their knowledge and expertise to this project. I am also indebted to Motion Reality Inc. for their vision and expertise in creating the MAT-T system, and for allowing me to train golf professionals all over the world on how to use it.

I've had several mentors in the golf industry without whom I wouldn't have been able to write this book. I want to thank teachers Mike Bennett, Andy Plummer, and Ben Doyle; the late, great Marlow Quick; and LPGA teaching professional Cathy Mant. I also want to thank my close golf colleagues—Jim Dunlap, Jason Owens, Adrian Burnter, Bryan Tunstill, "Big Daddy" Pash, Dominic Marconi, Kirk Nelson, and Bob Owan—and all of the staff at Pumpkin Ridge Golf Club. Special thanks to PGA Tour player Brian Henninger for all his support and friendship; my former college teammate at Brigham Young University-Hawaii, Dean Wilson, who I respect so much (he told us all that he would play on the PGA Tour one day, and he did); Champions Tour players Peter Jacobsen and Bruce Summerhays; and Dick Kramer, Jono Herrick, and Craig Griswold for giving me the opportunity to become a golf teaching professional.

Lastly, I want to thank my wonderful parents, who have never denied me anything in life, and my siblings Eddie, Wes, Smidge, and Boo, who always encouraged me to follow my dreams and let me beat them in golf. Most importantly, to my beautiful wife Janea and my five children, Samuel, Gabe, Charlotte, Lucy, and Henry: I love you so much and am incredibly grateful for all of the support you've given me throughout my career.